The World of Civil Aerospace

The World of Civil Aerospace

SAE INTERNATIONAL®

Warrendale, Pennsylvania, USA

400 Commonwealth Drive
Warrendale, PA 15096-0001 USA
E-mail: CustomerService@sae.org
Phone: 877-606-7323 (inside USA and Canada)
724-776-4970 (outside USA)
Fax: 724-776-0790

Library of Congress Catalog Number 2018948883
SAE Order Number R-468
http://dx.doi.org/10.4271/R-468

ISBN-Print 978-0-7680-9347-6
ISBN-PDF 978-0-7680-9553-1
ISBN-ePUB 978-0-7680-9349-0
ISBN-PRC 978-0-7680-9350-6
ISBN-HTML 978-0-7680-9348-3

To purchase bulk quantities, please contact: SAE Customer Service

E-mail: CustomerService@sae.org
Phone: 877-606-7323 (inside USA and Canada)
724-776-4970 (outside USA)
Fax: 724-776-0790

Visit the SAE International Bookstore at books.sae.org

contents

CHAPTER 1

Introduction: What We Don't See 1

References 6

CHAPTER 2

The Evolution of Civil Aerospace: Sometimes Crazy but True 9

The Early Days 9

Fixed Wing Flight 11

Commercialization 13

The Age of the Jet Engine 15

Bigger and Better 17

Going Supersonic 18

More of the Same 19

Service Contracts and the Growth of Data 21

References 22

CHAPTER 3

Designing the Aircraft: Pushing the Envelope 25

Introduction 25

What Are Our Passenger Expectations? 26

Airliner Categories 26

The Design and Development Process 27

Aircraft Requirements Capture 27

Customer Needs: Air Transport Market Research 28

Safety and Environmental Requirements 29

Aircraft Top Level Requirements (TLRs) 30

Conceptual Design: Determination of the Initial Layout 30

Working Together to Design the Aircraft 32

Design to Meet or Exceed Passenger Expectations 33
Design for Safety 33
Design to Reduce Passenger Costs, While Delivering Profits to Manufacturers and Airlines 33
DOC Elements 33
Reducing Fuel and Oil Costs and Commensurate CO_2 Reductions 34
Reducing Depreciation Costs 34
Reducing Maintenance Costs 35
Reducing Crew Costs 36
Design for Comfort and Communications 37
Design for Speed and Flight Frequencies 38
Design for Environmental Responsibility 38
Introduction of Possible New Technologies and Configurations 38
References 39

CHAPTER 4
The Planemakers: Manufacturing the Products 41
Powering Up a New Industry 42
Specialization Grows 42
Birth of a New Industry 43
Building for the Jet Age 46
Expanding Partnerships 47
Digitally Enabled Manufacturing 49
The Powerplant Revolution 53
Managing the Supply Chain 53
References 55

CHAPTER 5
Testing: What Works and What Doesn't 57
Static Tests 58
Complex Machinery 60
New Procedures 62
Why Flight Test? 64
References 67

CHAPTER 6

The Journey of Certification: Safety Is Paramount 69

Types of Certificates 70

Process of Obtaining the TC 71

Testing to Support Certification 73

Engine as an Example 75

References 78

CHAPTER 7

Operators: What Makes Airlines (Well) Airlines 81

Introduction 81

Passenger Journey 82

Business Strategy 83

- Route Networks 83
- Network Planning 83
- Fleet Strategy and Products 85

Finances 87

- Costs 88
- Fares and Pricing 88

The Operation 88

- Operations Center 89
- Reservations/Distribution Channels 89
- Airport Customer Service 89
- Security 90
- Aircraft Load Agents 90
- Maintenance 90
- Scheduling 90
- Dispatchers 92
- Meteorology 92
- Servicing 93
- In-Flight Services 93
- Flight Operations 93
- Corporate Safety and Security 93

Other Supporting Functions 93
When Things Go Wrong 94
The Human Part: Employees 94
The Human Part: Passengers 96
References 97

CHAPTER 8
Maintenance: A Necessary Evil? 101
Introduction 101
Safety and Compliance 102
Regulations 103
Maintenance Programs and Planning 103
Maintenance Evolution 106
Big Barriers to Entry 107
Business Models Are Complex 107
OEMs Expanding into MRO Services 108
Airlines and MROs 109
Reducing Costs with Innovation 109
Health Monitoring, Big Data, and Analytics 110
RFID 110
STCs 110
References 111

CHAPTER 9
Safety: Culturally Embedded 115
Safety 115
The Origins of Safety Assessments 116
The Introduction of Design Assurance 118
The Safety Assessment Process 119
Safety Requirements Development and Validation 121
Aircraft Functional Hazard Assessment 122
Preliminary Aircraft Safety Assessment 122
System Functional Hazard Assessment 123
Preliminary System Safety Assessment 124
Software and Hardware Development Processes 125

Safety Requirements Verification 125
System-Level Analyses 126
Common Cause Analyses 126
System Safety Assessment 127
Aircraft Safety Assessment 127
Safety Management and the Safety Life Cycle 128
Design and Certification 128
Production 129
Continued Airworthiness 129
Operating Environment 129
The Future of Safety 130
References 131

CHAPTER 10
The Environment: Emissions by Consensus 133
Introduction 133
Climate Change 134
Emission Reduction 136
Fuel Efficiency and Aircraft Design 136
Operations and Infrastructure 138
Economic Measures 139
Alternative Fuels 139
Aircraft Noise 142
Conclusion 144
References 145

CHAPTER 11
Standards: Bedrock to Build On 147
Introduction 147
The History of Standards and the Aerospace Industry 148
The Aerospace Standards Landscape: Early Standards Development Organizations 149
Value and Drivers 150
Standards and Conformity 151
Quality Management Systems 152

Standards and Innovation 152
Collaborative Endeavors for Standards Development 153
Dual-Use Standards and MilSpec Reform 154
Standards in Civil Aviation Regulations 155
Standards Use and Content Consumption: Past, Present, and Future 157
Conclusion 158
References 159

CHAPTER 12
The Future of Civil Aerospace: Onward and Upward 161
The Current State 162
Propulsion 166
Airframe 173
Materials and Manufacturing 179
Smart Aircraft 180
Economic Models 182
References 183

Appendix 187
About the Authors 189
Index 195

CHAPTER 1

Introduction: What We Don't See

Ian K. Jennions
Cranfield University

> *...There is no indispensable man*
>
> Saxon White Kessinger

It's a cold, dark, wet Monday morning at Chicago's O'Hare Airport. You're taking the shuttle to Boston for a meeting, leaving the family asleep as you creep out of the house, hoping to get back on the early flight and see the children before they go to bed. Boarding, coffee in hand, you welcome a few hours of reviewing the material for today and hope that there are no unexpected delays to your itinerary. You need not bother unduly, you would have to travel the same route once a week for 4 years to experience a canceled flight [1]. It's also many times safer than driving and, with the return flight being some 1700 miles long and costing around \$120, very cheap at about 14c/mile. Just a little over 100 years since the Wright brothers designed and flew the world's first aircraft [2], and you expect this kind of cost/performance whenever you fly, without thinking about the enormous achievements, compliant with aerospace rules and regulations, that make all of this possible. The broad goal of this book is to bring together the "big picture" view of this industry and, in doing so, highlight the notable successes and the future direction.

The true value of the aerospace sector to global Gross Domestic Product (GDP) is almost impossible to calculate. The International Air Transport Association (IATA), in its World Air Transport Statistics, cites airlines as supporting $2.7 trillion in economic activity in 2015 [3]. Market analysts quote the top 20 commercial aerospace companies as having revenues of $233 billion in 2015 [4], so the answer lies somewhere in between, but is huge. Forecasts are bullish, buoyed by growing passenger demand in Asia-Pacific and the Middle East in particular, and low oil prices (42% decrease in 2015 from 2014) driving good airline profitability. Passenger load factors are some 80%, and Airbus and Boeing between them have an order book of 12,400 aircraft, an estimated 9 year's production. As we further drill down into these numbers we must also ask: what is the safety record of the industry and what are the challenges in the years to come?

In 2015 3.6 billion people flew, 48% of the world's population. The International Civil Aviation Organization (ICAO) is a United Nations (UN) agency working with 191 member states and industry groups to reach consensus on international civil aviation standards in support of a safe, efficient, secure, economically sustainable, and environmentally responsible civil aviation sector [5]. It recorded that in 2015 there were 2.8 accidents per million departures for large civil aircraft [6]. With 33 million departures in that year, this equated to 92 accidents of which only 6 had fatalities. Further, two of these were Germanwings 9529 (pilot induced) and Metrojet 9268 (bomb), the only accidents in 2015 in Europe and the Middle East regions, respectively.

In contrast, according to the Association for Safe International Road Travel, car crashes account for 1.3M deaths per year worldwide [7]. While 90% of these are in low-to-middle income countries, there are still 37,000 deaths in the United States alone. This U.S. figure is nearly two orders of magnitude more than that for schedule commercial flights.

Set against this track record, the challenges being set are equally impressive:

- IATA, which represents 260 airlines (83% global air traffic), predicts 7 billion passengers by 2034, double the number in 2015 [8]. Significantly, China will overtake the United States for passengers/year by 2029 and India will move from ninth to third, displacing the United Kingdom. Such an explosion of numbers, particularly in relatively new markets, poses a problem for Air Traffic Management (ATM) and safety in general. While the stories we read in the newspapers are of (few) airline accidents, if the safety rate stays the same and the number of flights doubles, then there will be obviously twice as many accidents on our front pages.
- Flightpath 2050 [9] is Europe's vision for aviation for 2050 and contains a number of highly ambitious goals, including:
  - One accident per 10 million commercial airline flights, a substantial reduction over the numbers mentioned above. This will be achieved by reducing human error through new designs and training, and technologies that support decision making.
  - 90% of people door to door anywhere within Europe within 4 hours. Not only are various modes of transport seamlessly interconnected, but security

and immigration checks are made as passengers proceed to the aircraft in a non-obtrusive manner.

- Flight arrivals within 1 minute of schedule, regardless of weather. The transport system is resilient with forecasting, and alternatives, planned beforehand.
- 50% reduction in certification cost to enable new aircraft to enter service quicker. This involves streamlining the engineering process and creating a new generation of standards.

- European Union (EU) 216/2004 is a European regulation that sets out the rules for compensation for delayed or canceled (D&C) flights [10]. Payment for compensation is not limited to the price of the ticket, and so the airline can end up paying a passenger far more in compensation. The regulation is very convoluted in its detail [11], with how much is to be paid and for what length of delay, being complicated by where the flight originates and ends. Airlines are going to have to focus on D&Cs to avoid undue tolls.
- There is a race to keep up with ever-increasing passenger expectations, especially with the connected world [12]. Airlines are expecting enormous growth between 2016 and 2019 in internet and mobile phone access on board. Two-thirds of airlines will operate connected aircraft in the next 3 years. Already boarding passes are being issued to smartphones, and sometimes being re-issued as paper due to system handling problems! Automated bag check-in is catching up. It has been found that once passengers switch to electronic methods they do not switch back. A partial on-board answer to this problem is the advent of Bring Your Own Device, but more still needs to be done.

Having considered some of the top-level messages, now let's turn to a broad view of the aerospace sector. While the central actor is clearly the aircraft, there are several major supporting roles, as shown below in Figure 1. The aircraft needs to be designed, manufactured, and tested, with safety as a central tenet (enacted through certification). This process needs to be conscious of the environment, and is aided by the formation of industry standards. The development of the aircraft is usually thought of as taking place over some 4 years (launch customer to delivery), but in the case of both the B787 and A380, it took around 7 years, attributable to the complexity of modern aircraft and supply chains. The other elements in Figure 1 deal with the aircraft operation and maintenance in service, which takes place over a much longer time scale; it is not uncommon for an aircraft to still be in regular service 30 years after delivery. As the complexity of aircraft has increased so has the complexity of the maintenance required to keep it flying, e.g., the B747 and Maintenance Steering Group (MSG) documentation [13].

Many people spend their career in a specialized sector of this industry. They may be involved in marketing for an Original Equipment Manufacturer (OEM), ground maintenance for a Maintenance, Repair, and Overhaul (MRO) base, or a design engineer for an airframer. In each case they are mostly unaware of the intricacy of the fascinating segment in which they work and the interrelationships necessary to carry the flying public.

FIGURE 1 The Civil Aerospace Sector.

The aerospace sector is transitioning from a product- to a service-based economy, a move started by the engine manufacturers some 20 years ago. In a product-based business, income is derived from selling the product and subsequent spare part sales. This results in a large peak followed by future, uncertain, income. In a service-based business, income is derived from leasing the product and/or providing regular maintenance. This provides a much more even income over a long period of time and a deeper customer/provider relationship. Depicting the sectorial view given in Figure 1 as a value chain, in a simple service economy view, gives **Figure 2** [14]. What is shown here are the entities from Figure 1 and their interrelationships; only those at the top level are shown for clarity. The transactions shown are in one direction, as the return direction is, in all cases, money for the service provided.

There are two focal points in this figure, the Aircraft OEM and the Airline Operator. The Aircraft OEM designs, manufactures, and tests the aircraft, while abiding by statutory regulations. Financing of this venture may take many forms but Risk and Revenue Sharing Partnerships are probably the major vehicle. Here, suppliers to the OEM put up some of the capital needed to finance the new aircraft (not shown in Figure 2) in return for an agreement to manufacture parts. The suppliers therefore share the risk of the aircraft program, whether it's a success or not, and also the revenue. The other focal point is the Operator. They must deal with route planning, passenger numbers, regulation, and marketing, while keeping an eye on new aircraft entering the market and how to maintain those in their fleet. The service offerings in this area have brought new options to the Operator. Instead of going to an independent MRO for aircraft maintenance, some OEMs are providing MRO services themselves. Although not shown in

FIGURE 2 Aerospace value chain.

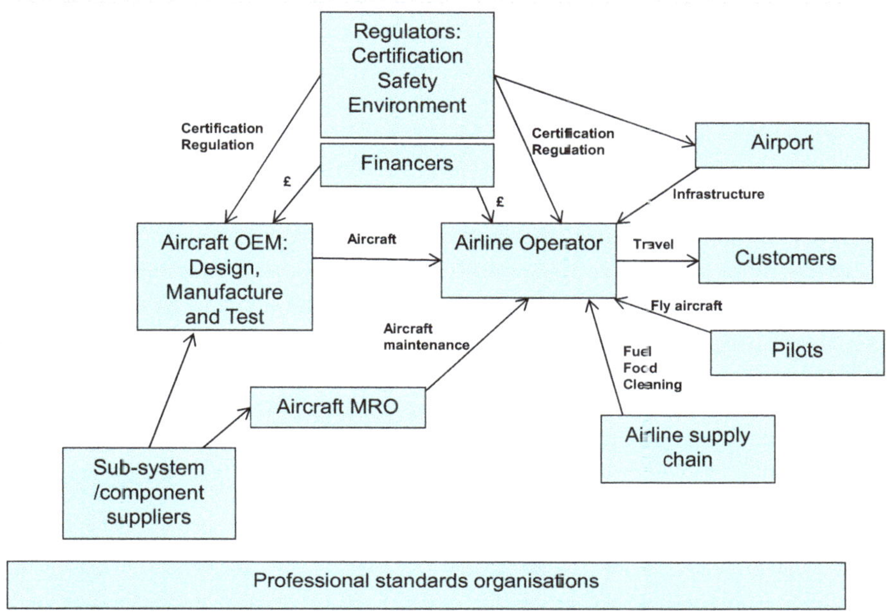

this diagram, for simplicity, the engine manufacturers have a lead in this field and have been offering maintenance, up to the level of guaranteeing availability, for some years now. Another option is for the Operators to create their own MROs, up to and including engine maintenance; this has been tackled by only a few Operators but is a very lucrative business. Other service offerings are apparent from Figure 2. For example, the parts suppliers to the OEM can supply spare parts to an independent MRO, but the OEM can aggregate these suppliers to provide a "one stop shop" for an outside MRO. Many more business opportunities exist in this field, but will not be covered in the book. Instead, it will concentrate on the existing businesses depicted in Figures 1 and 2.

The book is structured along the following lines and runs roughly clockwise from bottom left of Figure 1. It starts with a history of the sector, not of aircraft per se, but of their evolving design and the support that has had to go around it. From lessons such as the Comet [15], which changed aircraft design rules, through the maintenance restructuring that surrounded the Boeing 747, to the vastly more informated aircraft of today, the Airbus A380 and the Boeing 787.

It then examines how new aircraft are designed, including the market forces that drive new production and the integration of new technology onto these platforms. Included with this, but presented as separate chapters, are the manufacture and test of such complex pieces of engineering. Attention is then turned to how aircraft are operated, balancing capacity and load with small profit margins, and maintained, an area that continues to change rapidly with the introduction of health monitoring [16]. Being able to tell, before a part fails, that preventative maintenance is needed and can be scheduled so as not to disrupt flights is a step that is still being taken.

All of the chapters so far have described activities that have to be conducted with a view to certification and safety, the subject of the next two chapters. It has been estimated that certification costs up to 50% of the total development of a new aircraft, and the paperwork necessary to accompany this adds time to the development program. Alongside it, safety is the number one priority for civil aviation, and it has a good track record to prove this.

To round out our view of the industry, and complete the view shown in Figure 1, the environment and standards are addressed next. Concern for the environment has always been there but has grown considerably over the last 10-20 years. The chapter looks at ways in which environmental impact can be minimized, demonstrating another way in which the industry works collaboratively. Without standards all of the above work would be somewhat ad hoc; it is very important that the industry agrees to common standards and it has a good record of doing so. This represents mutually beneficial work by industry, the regulatory authorities, and others, for the benefit of all.

To conclude the book, a view of the future is taken, mentioning future technologies such as more electric/hybrid aircraft, supersonic flight, and solar power, to name but a few. One thing is for sure, if you look at the track record of aviation over the years, it's an exciting and, at times, unpredictable business to be in. *The World of Civil Aerospace* aims to make sense of it all.

References

1. Flight Statistics, http://www.flightstats.com/go/Media/stats.do.
2. Wright Brothers, https://en.wikipedia.org/wiki/Wright_brothers.
3. IATA WATS, 2015, http://www.iata.org/pressroom/pr/Pages/2016-07-05-01.aspx.
4. Deloitte's 2016 Aerospace and Defence Sector Outlook, https://www2.deloitte.com/global/en/pages/manufacturing/articles/global-a-and-d-outlook.html.
5. ICAO, http://www.icao.int/Pages/default.aspx.
6. ICAO Safety Report, http://www.icao.int/safety/Documents/ICAO_SR%202016_final_13July.pdf.
7. ASIRT, https://asirt.org/initiatives/informing-road-users/road-safety-facts/road-crash-statistics.
8. IATA Air Passenger Forecast, http://www.iata.org/pressroom/pr/Pages/2015-11-26-01.aspx.
9. Flightpath 2050, http://ec.europa.eu/transport/sites/transport/files/modes/air/doc/flightpath2050.pdf.
10. EU 261/2004 Explained, https://en.wikipedia.org/wiki/Regulation_261/2004.

11. EU 261/2004 from EU, http://eur-lex.europa.eu/legal-content/EN/ALL/?uri=celex%3A32004R0261.

12. SITA Airlines IT Trends Survey, https://www.sita.aero/resources/type/surveys-reports/airline-it-trends-survey-2016.

13. B747 and MSG, http://www.skybrary.aero/index.php/MSG-3.

14. Jennions, I.K. ed., *Integrated Vehicle Health Management: Business Case Theory and Practice*, (Warrendale, SAE International, 2013), ISBN 978-0-7680-7645-5.

15. Comet Accident, https://en.wikipedia.org/wiki/BOAC_Flight_781

16. Jennions, I.K. ed., *Integrated Vehicle Health Management: Perspectives on an Emerging Field*, (Warrendale: SAE International, 2011), ISBN 978-0-7680-6432-2.

CHAPTER 2

The Evolution of Civil Aerospace: Sometimes Crazy but True

Richard Greaves
S.V.P. Engineering, Meggitt Plc – Retired

For once you have tasted flight you will walk the earth with your eyes turned skyward, for there you have been and there you will long to return

Ascribed to Leonardo da Vinci (1452-1519)

The Early Days

It may well be, given the existence of birds, that humanity has thought of flight for many millennia. We can certainly go back to the Chinese in the fourth century BC when the use of kites was documented. The Chinese also exploited the fact that hot air rises, leading to the invention of floating lanterns.

The idea that a person may be able to fly by copying the actions of a bird was well embedded in philosophical thought by the time of the ancient Greeks, as seen in the story of Icarus and his attempt to fly with wings attached to his body. The concept of attaching wings to people and flying continued for centuries. The great Renaissance polymath

FIGURE 1 A Leonardo da Vinci sketch.

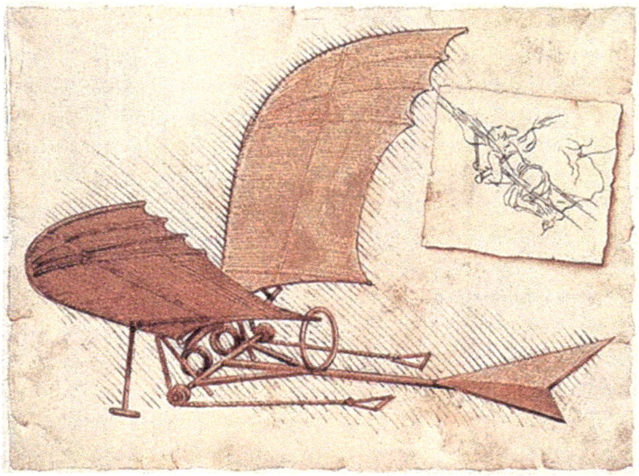

Leonardo da Vinci drew many conceptual drawings of flying machines, with flapping wings known as "ornithopters," c 1501; see Figure 1 [1]. Other sketches bear a resemblance to helicopters.

Serious development toward modern flying machines advanced once some of the basic laws of physics that had been established by Sir Isaac Newton [2]. Development could proceed on an understanding of the laws of physics. The first aviation publication followed in 1716, "Sketch of a machine for flying in the air," by the Swedish physicist and mathematician Emanuel Swedenborg [3].

The Englishman Sir George Cayley (1773-1857) may be regarded as one of the most important people in the history of aviation. In 1799 he designed a fixed-wing flying machine with separate systems for lift, propulsion, and control [4]. He correctly identified the four vector forces that influence an aircraft—thrust, lift, drag, and gravity—and in 1810 he published a three-part treatise called "On Aerial Navigation." Sketches in his notebook indicated that, as early as 1792, he understood the principle of a lift-generating inclined plane. In the 1840s he built a biplane glider which flew and, in 1853, a larger glider was flown on Cayley's country estate, Wydale Hall in Yorkshire.

During George Cayley's early years, successful experiments in lighter-than-air flight were being carried out across the English Channel in France. The first public demonstration of a hot-air balloon flight took place on June 4, 1783, at Annonay in France. The balloon was the invention of the brothers Jean-Michel (1740-1810) and Jacques-Etienne (1745-1798) Montgolfier. It weighed 500 kg with a volume of 790 m^3. It flew two kilometers; the flight lasting ten minutes. This soon led to the first (tethered) manned flight by Etienne Montgolfier which occurred in Paris in front of King Louis XVI on October 15, 1783. The first free flight took place on November 21, 1783, and went 9 kilometers and up to 910 m in altitude over Paris; it lasted 25 minutes.

FIGURE 2 Ascent of Mammoth balloon, May 28, 1863, with Henry Negretti on board.

Development of balloons, both hot air and lighter-than-air hydrogen and later helium, continued in the 19th (Figure 2) and early 20th centuries.

Negretti and Zambra [5] developed the world's first altimeter which they demonstrated at the Great Exhibition in London in the early 1850s when they went up in a hot-air balloon. This company was taken over in the 20th century and continues today; it builds altimeters as part of Meggitt Avionics in England.

The world's first aeronautical society, the Royal Aeronautical Society [6], was founded in Great Britain on January 1, 1866. This was the world's first aerospace professional association. It continues today and caters for all aspects of the aerospace community.

The forces of lift and gravity were overcome, but those of thrust and control were to prove more difficult. A source of thrust and its fuel had to be lifted into the air, and control and directional means added to the vehicles. This effort culminated in the achievements of Alberto Santos-Dumont (1873-1932) who on October 19, 1901, won the Deutsch de la Meurthe prize for flying round the Eiffel Tower in a dirigible from St-Cloud, covering a total distance of 11 kilometers in less than 30 minutes. Santos-Dumont then turned his talents to heavier-than-air machines, but in this he was beaten by the Wright brothers, in America. The commercial development of rigid airships was pioneered by Count Ferdinand von Zeppelin (1838-1917) in the late 19th century. On July 2, 1900, the first flight of his airship LZ-1 occurred from Friedrichshafen over Lake Constance lasting 20 minutes. Continuous improvements were made over the years, but on May 6, 1937, the Hindenburg disaster occurred when LZ-129 caught fire while attempting to dock in New Jersey. This resulted in the death of 35 of the 97 people on board and halted further airship development for many years.

Fixed Wing Flight

Fixed-wing aircraft development had continued apace since the first flight of the Wright brothers—Orville (1871-1948) and Wilbur (1867-1912). They made the first controlled sustained flight of a powered heavier-than-air aircraft on December 17, 1903, near Kitty Hawk, South Carolina.

They were the first to invent and demonstrate aircraft controls that enabled fixed-wing powered flights. Their invention was a 3-axis control which enabled the pilot to steer the aircraft and maintain its equilibrium. Their first U.S. patent number 821,393 [7] was not about claiming a flying machine, but rather a system of aerodynamic control that manipulated a flying machine's surfaces.

This was key to their success. Their rivals had been concentrating on propulsion (thrust). The Wright brothers owned a bicycle business in Dayton, Ohio, and were aware of the problem of balance when riding a bicycle and made the parallel with flying an aircraft. They also observed that birds changed the angle of the ends of their wings to make their bodies roll left or right. They discovered wing-warping and patented the idea. Through a system of wires and pulleys, they were able to twist the ends of the wings in flight, a function that is achieved on modern planes by ailerons.

In 1900, the Wright brothers went to Kitty Hawk and tested gliders that they had built. Over the next three years, intensive testing in Dayton and Kitty Hawk enabled them to finally build the Wright Flyer I (Figure 3).

FIGURE 3 The Wright Flyer flies December 17, 1903.

The Wright brothers even carved their own propellers. Not only did the aircraft have wing-warping but also an adjustable hinged rudder and a 12 hp engine that their mechanic built in six weeks. The engine cylinder block was cast aluminum—a rare thing at that time. The Wright Flyer cost $1000 to build (about $180,000 today), and a home-built wind tunnel was used to test fundamental aerodynamics. The Wright brothers continued to refine their aircraft, flying Wright Flyer II which achieved flights of over three miles in five minutes on Huffman Prairie, Dayton. Their Flyer III had independent controls for each of the three axes: pitch, roll, and yaw, leading to greatly improved stability. It achieved a 39.4 km flight in 38 minutes 3 seconds, safely landing when its fuel ran out. The Wrights incorporated the Wright Company on November 22, 1909—Wilbur was President and Orville Vice President. They henceforth spent most of their time selling their aircraft and defending their patents, especially against Glenn Curtiss. Wilbur, who died in 1912, took the leading role in the fight against Curtiss, who finally lost on appeal in January 1914. Orville felt vindicated by the decision and then sold the company in 1915. It is ironic that a successor, the Wright Aeronautical Corporation, merged with Curtiss Aeroplane in 1929 to form Curtiss-Wright.

The American Society of Aeronautic Engineers was founded on July 23, 1915, at the behest of Thomas A. Edison on behalf of the U.S. Navy Board. Founder members included Orville Wright, Glenn H. Curtiss, and Elmer A. Sperry to name a few. Discussion ensued to join the Society of Automobile Engineers which had been founded in 1905. Less than two years later, on January 11, 1917, at its annual general meeting, the Society of Automobile Engineers consolidated with the Society of Aeronautic Engineers to become SAE, The Society of Automotive Engineers. In those days "automotive" was applied to any vehicle on the ground or in the air that was self-propelling. The first international aerospace standard was produced in 1917—it was for interchangeable spark plugs for aircraft engines. Europe was at this time at war. The standard was developed with British, French, and Italian engineers. SAE is today the world's leading source of aerospace standards.

World War (WW) I gave an enormous impetus to the development of aircraft.

The roles that aircraft could play in the war were immediately perceived by the military on both sides, primarily reconnaissance and bombing. In 1915 the Germans fielded the Fokker monoplane (**Figure 4**).

FIGURE 4 Fokker Eindecker monoplane.

Anton Herman Gerard Fokker (1890-1939), a Dutchman, was inspired by Wilbur Wright's exhibition flights in Paris in 1908. He went to study mechanics in Germany and soon built his first plane, the Spider in 1910. He founded his own aircraft company near Berlin in 1912. Although the Fokker Eindecker [8] was not the world's first monoplane, that was the 1907 Demoiselle from Santos-Dumont, it was the most successful monoplane in WWI. Fokker is credited with the first application of the synchronization gear which enabled a machine gun to fire forward through the propeller without hitting the propeller blades.

The aircraft which is said by many to have won the aerial war on the Western Front for the Allies was the Sopwith Camel (**Figure 5**).

The Sopwith Aviation Company was founded in June 1912 by Thomas Octave Murdoch Sopwith (1888-1989) with Fred Sigrist and others. During the war, 5,747 Sopwith Camel aircraft were produced. It was introduced to the front in June 1917. It is credited with shooting down 1,294 enemy aircraft, more than any other allied aircraft in the war. It was said to be difficult to fly, but once mastered was the most agile fighter aircraft there was. It was equipped with two Vickers machine guns and could carry four bombs.

After the war, Sopwith Aviation went bankrupt. With his old friend and partner Harry Hawker, Thomas Sopwith became the founder and chairman of Hawker Aircraft in 1920. The company then played a key role in WWII with the Hawker Hurricane fighter and the Lancaster bomber.

FIGURE 5 The Sopwith Camel biplane.

Commercialization

During the early 20th century, other significant events in aviation were taking place led by famous pioneers in the field. We will start by considering William (Bill) Edward Boeing (1881-1956) who was born in Detroit, Michigan, of German immigrants. His father Wilhelm from Germany and his mother Marie Ortmann from Vienna arrived in the United States in 1868. After the death of his father, Bill attended school in Vevey, Switzerland, and Yale. He entered the timber business and moved to the Pacific Northwest. He grew fascinated by aviation and became a pilot, having attended the Glenn Martin Flying School in Los Angeles. In 1912, Glenn Martin started his own company making aircraft. This is now part of the Lockheed Martin Corporation. Bill Boeing, having bought a Martin aircraft, returned to Seattle. With his friend Cdr. George Westervelt of the U.S. Navy, they decided to build a better plane, which they did—an amphibious biplane which had excellent performance. In 1916, Boeing and Westervelt established the Pacific Aero Products Co. which in May 1917 became the Boeing Airplane Company—a world leader in the industry. In 1929 the United Aircraft and Transportation Company (UA&TC) was formed when Bill Boeing teamed up with Frederick Rentschler of Pratt and Whitney (P&W). This holding company soon brought in other partners and controlled the stock not only of the Boeing Airplane Company and P&W Aircraft, but also of the Chance Vought Corporation, the Hamilton Aero Manufacturing Company, the Sikorsky Aviation Corporation, Northrop Aircraft, the Stearman Aircraft Company, and the Standard Steel Propeller Company together with three young airlines and two airport companies, one in Connecticut and one in California. The big business at the time was delivering mail. The P&W Wasp-powered Boeing 40 was the airmail aircraft of the time. This led in 1933 to the first flight of the Boeing 247, the world's first "modern" aircraft—low wing, all metal construction, with retractable landing gear. In 1934, the U.S. government accused UA&TC of monopolistic practices and forced the company to separate flight operations from development and manufacturing. United Airlines was spun off. Rentschler created United Aircraft Corporation (now United Technologies) which included P&W, Hamilton Standard, and Vought/Sikorsky. Boeing/Northrop became its own company.

A further pioneer who influenced today's aerospace industry was Igor Ivanovich Sikorsky (1899-1972). Born in Kiev, in 1908 he went to Germany with his father and learned of the exploits of the Wright brothers and Ferdinand von Zeppelin. After engineering school, he set up his own company in Kiev, manufacturing aircraft. He experimented with helicopter design in 1909 but determined that the technology of the day would not enable him to realize a working prototype. In 1912, Sikorsky became chief engineer of the Russian Baltic Railroad Car Works in St. Petersburg. He designed and built a four-engine aircraft, the Russky Vityaz, the first four-engine aircraft in the world. This led to the Ilya Muromets (Sikorsky S-22) which, from June 30 to July 12, 1914, established a world record by flying from St. Petersburg to Kiev and back (1,200 km). It could carry 16 passengers in comfort. At the outbreak of WWI, this aircraft became a bomber, the Ilya Muromets type V, the world's first purpose-built bomber.

In 1918, after the Communist Revolution in Russia, having been warned that the Bolsheviks would shoot him, he fled the country with his family to France—then continued to the United States, arriving in 1919. He founded the Sikorsky Aircraft Company in 1923 and in the 1930s was doing a strong business building flying boats. When the era of flying boats faded, Sikorsky returned to his first love: helicopters (**Figure 6**). By now technology had advanced to the point where he thought that he could build a working helicopter. In 1939, the Sikorsky VS-300 made its first flight. It was the world's first practical, working helicopter. Forty years had gone by, but now technology had caught up with his vision a common theme throughout aviation history.

FIGURE 6 Igor Sikorsky and his helicopter.

From envisaging human-powered flight by flapping wings, to hot air- and hydrogen- and later helium-powered balloons and dirigibles, the internal combustion engine had come to dominate flight in the early 20th century. The form of the engine was rotary (**Figure 7**). An odd number of cylinders (usually 7 or 9) were placed in a row in a radial configuration.

The entire crankcase and attached cylinders rotated around the stationary crankshaft. The number of cylinders needed to be an odd number to maintain smooth running—every other piston being the firing order. The engines ran smoothly, the large rotating mass acted as a flywheel. The propeller was simply bolted onto the front of the crankcase.

After WWI, radial engines which were "stationary" in that the crankcase and cylinders that did not rotate with the propeller came to dominate the engine type, not having the inertia problems associated with the rotary.

FIGURE 7 The Le Rhône 9C, typical rotary engine of WWI.

The Age of the Jet Engine

The person who came up with the technological breakthrough that led the world into the jet age was Sir Frank Whittle (1907-1996), who was born in Coventry, England. Young Whittle was brought up in a mechanical engineering environment. He developed tremendous practical mechanical ability, was a first-class (rather daredevil) pilot in the Royal Air Force, retiring as Air Commodore, and obtained a First Class Honours degree in Engineering from Cambridge University.

Whittle's breakthrough came when thinking about how to overcome the inherent limitations of the rotary engine and may be stated as "why not substitute a turbine for a piston engine?" Instead of using a piston engine to produce compressed air to the combustor (burner), a turbine could be used to extract some power from the exhaust to drive a compressor, and the remaining exhaust thrust would power the aircraft. Whittle sent his concept to the Air Ministry in 1929, but it was returned with the comment that his idea was "impracticable." He was persuaded by his friend Pat Johnson to patent the idea, which he did in 1930. After finishing his university studies, Whittle started to raise funding to set up his own company to demonstrate his idea. Power Jets Ltd was established on January 27, 1936, and development started in Rugby in the facilities of British Thomson-Houston. The first Whittle Unit ran on April 12, 1937—this is conventionally regarded as the invention date of the jet engine (Figure 8). The British Government still showed little interest in the invention until 1939

FIGURE 8 Sir Frank Whittle and his W1 Engine.

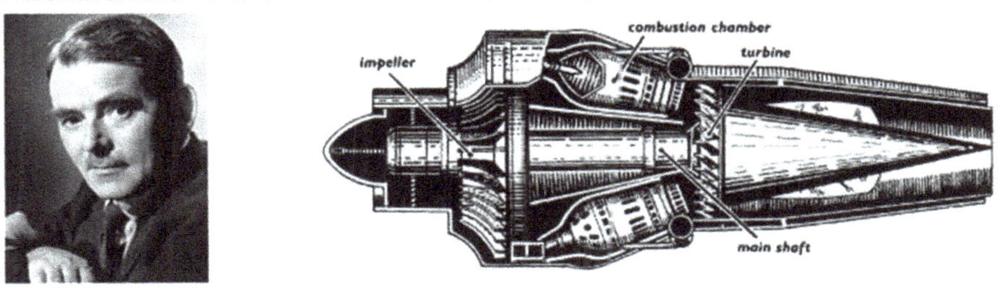

when the Germans flew the first aircraft powered by a jet engine on August 27, 1939. Hans Joachim Pabst von Ohain (1911-1998), a German physicist, had independently invented the jet engine. The German government immediately saw its possibilities and flew it in a Heinkel H178 aircraft. World War II broke out a week later. This woke up the British government but the first jet-powered flight in the United Kingdom did not take place until nearly two years after the Germans', on May 15, 1941, when a Gloster E28/29 was fitted with a Whittle W1 Engine.

The breakthrough for Frank Whittle and Power Jets came in 1940, when at the prompting of Stanley Hooker, Ernest Hives the chairman of Rolls-Royce visited Power Jets. He was so impressed that he offered that Rolls-Royce would make any parts that Whittle wanted in their Derby facility. This started Rolls-Royce on the path to become one of the world's leading developers and manufacturers of jet engines. During WWII, under agreement between the United Kingdom and the United States, Whittle's technology was transferred to the United States. A production license was granted to General Electric in September 1941. Following improvements, the 1-40 (J33) engine, which powered the Lockheed P80 Shooting Star fighter aircraft, went into operation in 1944. P&W followed with jet engine production in the early 1950s. Both also became world leading companies in the field of jet engine development and production.

The last 70 years have seen the aerospace industry both commercial and military dominated by the jet engine as the means of propulsion. The first purpose-built jet airliner was the British de Havilland Comet which first flew in 1949 and entered service in 1952. With a turbojet engine the air passes through the turbine and the combustion chamber. This design mutated in the 1970s to the turbofan engine (**Figure 9**). In the design a ducted fan uses the mechanical energy from the gas turbine to accelerate air

FIGURE 9 Turbofan engine.

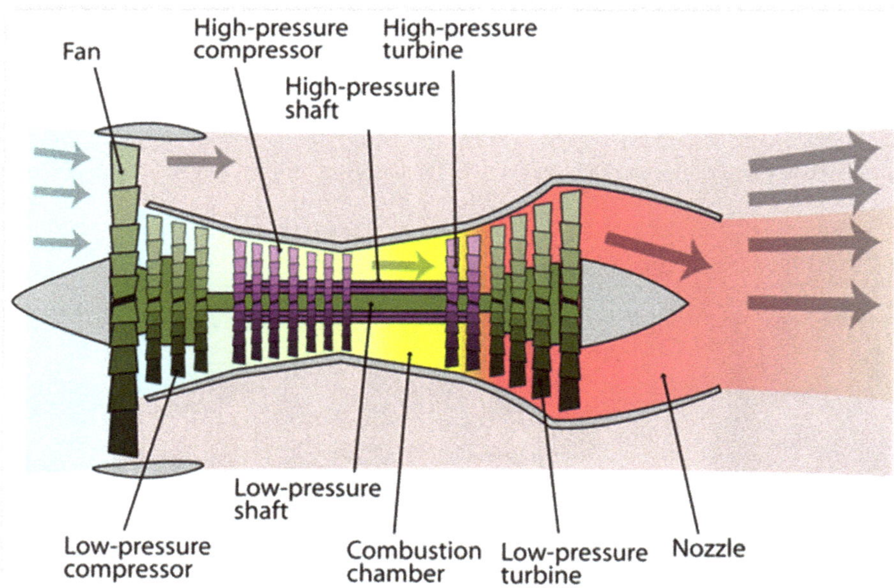

K. Aainsqatsi/CC BY 2.5.

rearwards to produce thrust, while a smaller portion of the air is directed into the "core" to be compressed, mixed with fuel, and burnt in the combustion chamber to generate the energy to drive the turbines. The basic principle of the design (for a two-spool jet engine) is shown below. This is the general design for GE and P&W engines, while Rolls-Royce mainly uses a three-spool design with an intermediate pressure stage, to increase overall efficiency. Recently, P&W has commercialized a geared turbofan design that aims to achieve higher efficiencies without adding the mechanical complexity of a third spool.

Refinement of the design continues today, the essential drivers being reduction in emissions, reduction in noise, and reduction in fuel burn. Each new aircraft design requires specific engine capability to optimize overall configuration.

After WWII commercial aviation grew rapidly. The Convention on International Civil Aviation was signed in Chicago on December 7, 1944, by 52 countries. The signatories agreed to harmonize and standardize the use of airspace for safety, efficiency, and regularity of air transport. The International Civil Aviation Organization (ICAO), a United Nations agency, was established on April 4, 1947, with its headquarters in Montreal, Canada. As of November 2017, there are 192 members. In effect, ICAO is the regulatory authority for civil aerospace worldwide.

The first commercial jet airliner to fly was the British de Havilland Comet in 1952, powered by four de Havilland Ghost turbojet engines, which were later replaced by four Rolls-Royce Avon engines. The corners of the square windows were stress concentrations, causing metal fatigue that led to a catastrophic failure of the fuselage. After three aircraft broke up in flight, the aircraft was withdrawn from service. It took the engineering community a number of years to understand the fundamentals of why this had happened.

By 1958, Boeing's first jet airliner, the B707, entered service. The Douglas Corporation had long led the civil air transport market with its series of piston engine DC aircraft during the 1920s, 1930s, and 1940s. The DC-3 was tremendously successful, a total of 16,079 having been built in all it variants. Donald Douglas was not convinced of the transition from piston engines to jet engines, but, with the advent of the B707, the Douglas Corporation needed to develop a competitor, which it did. The Douglas DC-8 (also powered by four engines) entered into service in 1959. In 1960, the Convair division of General Dynamics introduced into service their competitive aircraft, the Convair 880.

Bigger and Better

There was significant increase in air passenger travel in the 1960s. This led companies to think of designing larger aircraft. The next generation of aircraft, a truly visionary effort, was pioneered by Boeing, the wide-bodied (twin aisle) aircraft with the designation B747. This four-engine double-decker aircraft first flew in 1970. It was anticipated at the time that supersonic flight would take over from subsonic, so it was initially planned to only build 400 aircraft. But this did not transpire, and for 37 years it was the largest passenger aircraft in service, and by August 2017, 1,533 aircraft had been built. The development of this aircraft was made possible by several notable achievements. Boeing did not have a facility big enough for the 747, so a new manufacturing plant in Everett, Washington, USA, was built. This is the largest building, by volume, ever built. To power the aircraft, P&W developed the first high bypass jet engine, the JT9D. To meet safety and handling concerns, multiple system redundancies were incorporated for hydraulics, landing gear, and control surfaces. Also, the whole question of maintenance had to be re-examined. Too much maintenance can in fact induce failure and, on such a large complex vehicle, would have rendered its operation economically non-viable. Boeing helped establish an industry consortium called the Maintenance Steering Group (MSG) within the Air Transport Association (ATA) which developed a set of maintenance guidelines for maintaining the B747. This led to MSG1 and the concept of Reliability-Centered Maintenance (RCM). A seminal report by Nowlan and Heap [9] was produced by United Airlines, an ATA member, under contract to the Department of Defense. This report showed that 80% of failures are random; many scheduled maintenance tasks are irrelevant, there is not a "right" time to overhaul, and a minority of failures are age related.

Boeing was not left alone in the market place of large wide-bodied aircraft. The MacDonnell Douglas DC-10 three-engine aircraft entered into service in 1971. The Lockheed L-1011, also with three engines, entered into service in 1972 but was a commercial failure. After production ended, Lockheed withdrew from the commercial aviation market altogether.

While this was going on in the United States, European companies such as Hawker Siddeley, British Aircraft Corporation, Nord Aviation, and Breguet Aviation realized that they could no longer go it alone to design and manufacture civil air-transport aircraft, which could compete with the American manufacturers. A collaborative agreement between the United Kingdom, France, and West Germany was made in 1967. The British withdrew government support in 1969, and a new agreement between Germany and France created Airbus Industrie. The British company, Hawker Siddeley, continued as a commercial partner and was contracted to design and build the wings. All subsequent Airbus aircraft have had UK-built wings. Airbus developed the world's first wide-bodied twin-engine aircraft, the A300B, which was launched in 1974.

Going Supersonic

While the Americans were developing wide-body civil transport aircraft, British Aircraft Corporation (and Aerospatiale in France through leaked copies of BAC designs) came up with a concept for a narrow-body delta "ogee" wing supersonic transport (SST) aircraft. The French and UK governments signed an agreement in 1962 to go ahead with a joint project called the Concorde (**Figure 10**). First flight was in 1969 and the aircraft entered service in 1976. Concorde had a maximum cruise altitude of 60,000 ft, although it normally cruised at 56,000 ft at an average cruise speed of Mach 2.02. In the end, a total of only 14 aircraft entered service with British Airways and Air France—a technical masterpiece but a commercial failure. The aircraft was retired from service in 2003, three years after a fatal take-off crash in Paris and following a downturn in premium Transatlantic air travel after the 9/11 attacks in New York.

In America, Boeing won a competition from the U.S. government to design an SST aircraft. The Boeing design, the B2707, beat the Lockheed L-2000 design; contract award went to Boeing in January 1967. However, opposition concerning noise, sonic booms, depletion of the ozone layer, and commercial feasibility led to the abandonment of the project in 1971. Due to this, and a downturn in the aerospace industry, Boeing laid off 60,000 workers in Seattle. A billboard was erected near Sea Tac Airport in 1971 which read: "Will the last person leaving SEATTLE – Turn off the lights" [10].

FIGURE 10 Concorde landing at Farnborough, September 1974.

More of the Same

Aircraft design in the Soviet Union also flourished in the post–WWII period, led essentially by Tupolev and Ilyushin. To a large extent, their aircraft were following Western designs, such as the Tu154 resembling the B727, the Tu144 like the Concorde, and the Il 86 like a mini B747. Because of the Cold War, these aircraft mainly operated in the USSR and certain developing nations. They resembled Western designs visually but developed a poor reputation for reliability, and their engines were uncompetitive with Western jets, which explained their lack of appeal in Western markets.

Rolls-Royce in the United Kingdom was one of the "Big Three" Western manufacturers of advanced high-thrust turbofan engines, alongside P&W and General Electric in the United States, and since the 1970s these three companies have dominated the civil market for powering wide-body transport aircraft. The adoption of new materials, including carbon and a mix of carbon and titanium fan blades, have given these massive engines (with over 100,000 lbs of thrust) enough power to allow twin-engine wide-body aircraft to replace the classic four-engine format for even the longest international flight sectors—a development that was unthinkable for safety reasons even two decades ago. The huge 600+ seat Airbus A380 (Figure 11) has suffered badly by having four engines in competition with stretched twin-engine designs such as the super-efficient Boeing 777 and Airbus A350, both of which can carry up to 400 passengers. The manufacturers of large wide-body airliners have consolidated around just two suppliers, Boeing and Airbus, although Russia and China are now cooperating on a new generation of wide-body commercial aircraft for service in the second

FIGURE 11 A380 on final approach.

half of the next decade. It remains to be seen if these will have any appeal beyond their own domestic markets in the face of such a well-established global support infrastructure enjoyed by Boeing and Airbus.

Perhaps the most surprising development since the 1970s has been the extent to which the commercial market has remained wedded to the concept of a conventional 150-seat solution for short- and medium-haul air services. The classic design featuring twin underslung engines, on a low-sweep wing and narrow fuselage, has remained the most popular shape in the skies with airlines around the world. The aircraft that introduced the first product, the Boeing 737, has been upgraded and improved over 50 years, and to date no less than 10,000 have been built, with production still running at around 50 each month. Its arch rival, Airbus, has already delivered over 8,000 of its A320 aircraft in its first 30 years, and had a sales backlog of more than 6,000 yet to deliver at the time of this writing. The reason why these two aircraft have remained so competitive and popular is because they have incorporated, as new engines, avionics, and aerodynamic improvements have become available, sufficient performance gains to justify investment in low-risk new models where there is already a large fund of operating experience and a ready supply of spares and training facilities. Fitting advanced wingtip extensions is just one example where relatively simple improvements can offer a worthwhile enhanced performance, translating into reduced fuel bills and extended endurance. It all adds up to low risk and increased profits. While both Boeing and Airbus have continued to investigate and explore more radical next-generation designs, the paying customer has demanded more of the same. Boeing's exciting Sonic Cruiser in the late 1990s was a truly disruptive proposal for a high-speed subsonic long-range airliner, but the high operating costs and lack of evidence supporting a positive business case led airlines to ask for a better conventional airliner instead, exploiting more composite materials and new engines, and that led to the B787 design. This in turn forced Airbus to react by offering the A350, with a similar hi-tech specification.

But sharing the same familiar configuration are scaled down twinjets seating typically 110-140 seats, and which have also sold in their thousands, built by Brazil's Embraer, Canada's Bombardier, the Russian/Italian Sukhoi Superjet, and Japan's Mitsubishi. With the development of a new range of lightweight turbofans from CFM (the LEAP model) and P&W (the Geared TurboFan, GTF), a strong sales battle has emerged for future orders for the latest versions from these manufacturers. Embraer has thoroughly modernized its EMB 170-195 regional jet family into the E Series (**Figure 12**), which can carry up to a maximum of 140 passengers, below the economical operating optimum size of the B737 and A320. Bombardier's new C Series is also in this same size bracket and competition is very great. A Boeing-sponsored legal challenge over alleged dumping of C Series aircraft at a low price, which sought to impose crippling surcharges on C Series sales to U.S. airlines, was defeated and the Canadian supplier in its own defense pointed out that there was not a direct Boeing product in this size category. This case highlighted the high stakes involved in the most heavily contested sectors within the commercial market.

FIGURE 12 Embraer E series.

Many thought in the 1980s and 1990s that the availability of 50-70-seat twinjets for regional services from Embraer and Bombardier would bring to an end the reign of the twin turboprop airliners, such as the Bombardier Dash Q400 and the ATR family. However, this has not happened, and while sales of small regional jets have shrunk considerably, turboprop deliveries are on the rise, as they continue to offer operators a very economical solution on short-haul flights, especially using short runways or at airports with challenging local conditions. The appearance of small turboprop commercial aircraft has changed little since the 1950s, but this is misleading as today's products not only feature very fuel-efficient engines, with low maintenance needs, but their cockpits are as modern as in the largest wide-body jets, with glass displays and the latest avionics and navigational aids. Some even have Head-Up Displays giving crews maximum situational awareness, which is particularly valuable when flying into small airfields, perhaps surrounded by mountains or close to the sea, and where bad weather is an ongoing feature of operations.

Service Contracts and the Growth of Data

The amount of data being produced by the latest aircraft in service, such as the A380, A350, and B787, is enormous and rapidly increasing. It is estimated that by 2026, 98 million terabytes of data will be produced per year. Yet, air transport has long been one of the safest modes of transport. Worldwide fatalities in the period 2007-2016 were 2,832 passengers and 104 people on the ground, giving a total accident rate of less than one per million-flight departures. Most of the deaths were caused by loss of control in flight, followed by controlled flight into terrain. 2017 was the safest year ever in the history of civil air transport. There were no fatalities regarding FAA Part 25 Transport Category Aircraft anywhere in the world.

The data now available on modern aircraft may be said to be drowning us. However, this data enables the condition or health of the aircraft to be known in real time. Knowing health in real time is useful, but of more use is prognosis. How long can safe operation of the vehicle continue? This is a current area of much research and development. However, Aircraft Health Management (AHM) has now advanced to the point where some of its aspects are being adopted by the MSG-3 group to augment maintenance practices—the key high-level objective being, according to the A4A MSG working group, to "*determine means where analysis of data acquired from AHM systems can be used to define, improve or allow an alternative to a scheduled maintenance task.*"

The availability of so much real-time data flowing from air operations globally has become one of the most important sources of new business for the OEMs and other players in the maintenance and repair sector. Over the lifetime of a commercial aircraft, which might be between 20 and 30 years, the additional business and profit that can be generated by providing a comprehensive package of support services can far outweigh the original sales contract for the fleet buy. This aspect of commercial air transport management is rapidly changing to exploit the new electronic support products and associated technologies that allow greater flexibility for operators, with a reduced demand for very expensive in-house engineering facilities and skilled training facilities.

Support packages can be tailored to suit individual airline needs, with guarantees that spare parts or emergency repairs can be provided as and when required, and at short notice, without the airline having to anticipate, order, store, and transport the key parts, and have the skilled staff available to do the job without delay. The work is done for them as needed. The old pattern of providing all this support capability in-house in your own maintenance centers, and having it available on a stand-by basis, is increasingly unaffordable, and also a great drain on operating profitability. If support can be provided on demand within a contracted package, then new businesses can be created to do this work, enjoying all the advantages of economies of scale in procurement, spares holdings, and provision of skilled personnel. Airline support becomes a profit center rather than a financial burden. Such package solutions have long been a major sales feature promoted by Rolls-Royce in its TotalCare® product, and this is being further developed into a more integrated support entity known as the Intelligent Engine. Boeing is massively expanding its investment in its GoldCare support business, providing complete maintenance and 24/7 technical support for a growing number of airlines who are introducing large new fleets of aircraft and who see this approach as a key advantage in keeping support costs affordable. Airbus is following suit, as are other major aircraft manufacturers. One of the biggest challenges facing the commercial aviation sector is how to find and train the hundreds of thousands of next-generation engineers needed to keep tomorrow's airline fleets flying. The answer must be to adopt the best technologies and management systems, including artificial intelligence in the next few years, to make more support activities automated or robot-assisted. Examples can already be seen where advantage is being taken to use small camera-equipped drones for the inspection of very large aircraft moving surfaces. Such inspections can be carried out in a fraction of the time taken using inspection towers and overhead walkways. Other routine activities are also very suitable for increased computer-controlled operation.

With all the development around the world, it is clear that the growth of the aviation industry is far from over. New types of aircraft, such as electric aircraft, new types of engines, such as the geared turbo-fan, and new types of business models, such as air-taxi services, will ensure that this remains a healthy and expanding sector for many decades to come.

References

1. Sir Isaac Newton, "Philosophiae Naturalis Principia Mathematica," 1687, https://plato.stanford.edu/entries/newton-principia/.
2. Swedenborg, E. and Polhem, C., "Daedalus Hyperboreus," 1716.
3. Cayley's Flying Machine, https://www.uh.edu/engines/epi1120.htm.
4. Negretti & Zamba, "Instrument Makers," http://www.barometerworld.co.uk/negrettiandzambra/index.htm.

5. Royal Aeronautical Society, https://www.aerosociety.com/.
6. http://invention.psychology.msstate.edu/i/Wrights/WrightUSPatent/WrightPatent.html.
7. Fokker Eindecker, https://www.windsockdatafilespecials.co.uk/fokker-eindecker-compendium-1-660-p.asp.
8. Nowlan and Heap, AD-A066-579, 1978.
9. Evaluation Criteria for Reliability-Centered Maintenance (Rcm) Processes, JA1011_199908, 1999, https://www.sae.org/standards/content/ja1011_199908/.
10. A Guide to the Reliability-Centered Maintenance (Rcm) Standard, JA1012_200201, 2002, https://www.sae.org/standards/content/ja1012_200201/.
11. http://www.historylink.org/File/1287.
12. Leonardo da Vinci Ornithopter, https://science.howstuffworks.com/transport/flight/classic/ten-bungled-flight-attempt2.htm.

CHAPTER 3

Designing the Aircraft: Pushing the Envelope

John P. Fielding
Cranfield University

Science is the study of what is. Engineering builds what will be. The scientist merely explores that which exists, while the engineer creates what has never existed before.

Theodore von Karman

Introduction

Aircraft design is a complex, creative, iterative activity, which affects all aspects of the aircraft's life cycle. All chapters of this book are affected by aircraft design, but this one will concentrate on the various phases of the design process.

It starts with the ultimate customer, the passenger and his or her expectations of their journey. Next, a description is given of airliner categories and a summary of the design process, starting with the capture of passengers and other stakeholder requirements. A description will then be given of the conceptual design phase. Close collaboration is required between design, manufacturing, suppliers, and airline organizations, and this will be described. Reverting to passenger expectations, methods of meeting, or exceeding them are then explored in more detail. The chapter concludes with suggestions for exciting new technologies and aircraft configurations.

What Are Our Passenger Expectations?

There are as many passenger expectations as there are passengers, but the main ones are summarized here, and how to meet them will be discussed further on. The prime requirement is that the aircraft must have extremely high levels of **safety**, which can only be achieved by rigorous attention during all phases of its life cycle, from design through manufacture to operations.

Ticket **cost** is a major consideration for passengers, whether they book themselves through an agent or are paid for by their employer or sponsor. Minimum cost is crucial, but this must be balanced against other priorities listed in this section.

An important passenger priority is that of **comfort**, which is affected by personal space, ride quality, seat comfort, vibration, noise, temperature, and humidity. Passengers usually require meal and drink services and luggage stowage. Some airlines usually only provide these at extra cost, but have not yet made charges for using the washroom!

Productivity and enjoyment require good **communications** which are provided by in-flight entertainment, Wi-Fi, telecommunication, and computer connectivity. The main motivation for flying has always been time saving, relative to other transport modes. The total journey time is important, as is having sufficient flexibility in terms of travel dates and times, thus **speed and flight frequencies** are crucial.

Most people are aware of our planet's finite resources and humankind's responsibility for anthropogenic climate change. Some passengers prefer to fly in "greener," quieter aircraft, or to offset their flight pollution emissions. Therefore, the aerospace designer has significant **environmental responsibilities**.

Airliner Categories

There is a wide spectrum of categories of passenger airliners, starting with short-range **Regional Airliners**. The smallest have a capacity of 19 passengers and are usually powered by turboprop engines. They fly at some 300 mph over ranges of up to 1000 miles. Larger aircraft have similar performance, but carry up to 70 passengers. More recently, regional jets have been developed, with capacities between 35 and 130, having cruise speeds of some Mach 0.7. Regional aircraft can use shorter runways, and are therefore capable of using many more airports than larger types.

Single-Aisle Airliners have passenger capacities ranging from 130 to 220 seats, and usually operate over short-haul routes up to 3,000 miles, with the average of about 1,000. They cruise near Mach 0.75 and are the most numerous class of airliners, being used by many "Low Cost Carriers" (LCCs). Some LCCs are now operating over longer-range routes, such as the North Atlantic. **Small Wide-Body Airliners** operate over short, mid-range (4,000-7,000 miles) or long-haul routes (7,000-11,000 miles). They fly slightly faster, at some Mach 0.85, and utilize twin-aisle fuselage configurations.

Medium/Large Wide-Body Airliners are often developed as part of the same families as small wide-bodies and usually operate on high-capacity routes, over longer ranges. To increase capacity, the largest aircraft such as the Airbus A380 and Boeing 747 utilize full or partial upper-deck cabins.

The Design and Development Process

Figure 1 shows the widely accepted V diagram visualization of the entire life cycle of a complex product, such as an aircraft. It starts with the study of its concept of operation, which leads to the capture of requirements (shown below). It then continues through the design, testing, and operational phases of the aircraft.

Figure 2 amplifies the design phase into requirement capture and the conceptual, preliminary, and detail design stages. It illustrates the iterative nature of the design process, where a converging iterative spiral of design stages ends in the detail design, and ultimately manufacture and operation of the aircraft. Simplified semi-empirical methods are initially used. These are then progressively improved by the use of more accurate methods such as Computational Fluid Dynamics, Finite Element Structural Analysis, and Computer-Aided Design. Subsequent stages include wind-tunnel and structural testing and flight test. This process is usually successful if the spiral is convergent, but on occasions where the spiral is divergent, the design must be abandoned or subject to major modifications.

Aircraft Requirements Capture

Designing a new aircraft requires great commitment and the spending of huge financial and other resources. It is vitally important to start the process by deriving the right requirements. Aviation history has shown the results of failing to do this and the production of unsuccessful projects (Fielding, [3], Chapter 11).

FIGURE 1 The classic life cycle V diagram [1]

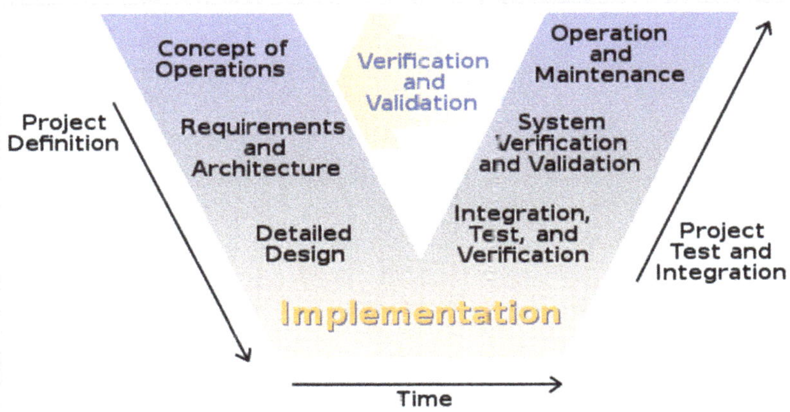

FIGURE 2 An illustration of an aircraft design process [2].

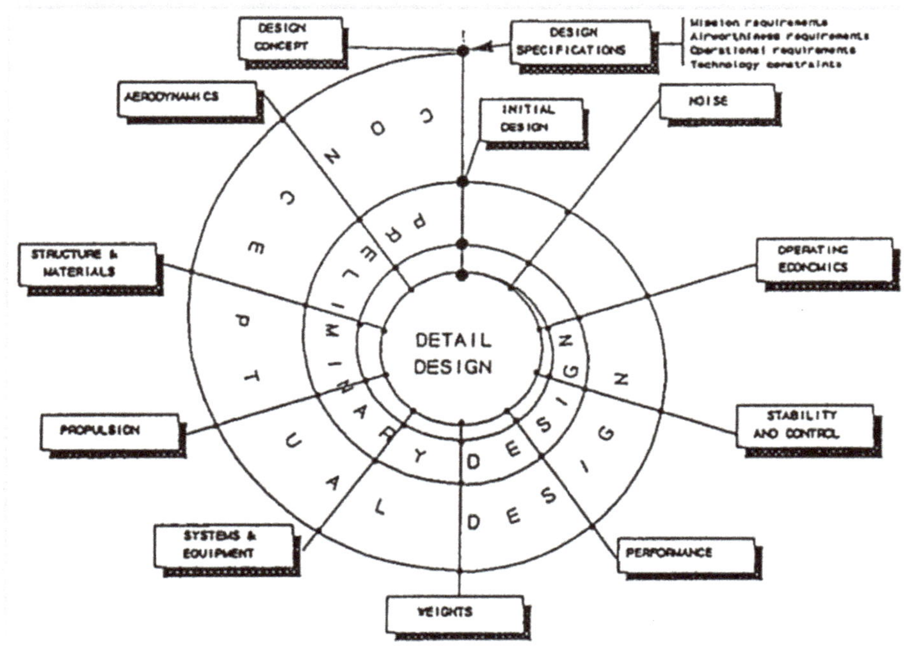

Aircraft requirements are derived from a combination of customer needs, passenger priorities, airworthiness standards, environmental requirements, company expertise, available technology, and non-available technology. The latter may be used if it becomes available at the right time in the program because of successful applied research and development processes.

Customer Needs: Air Transport Market Research: All companies employ marketing personnel to study future markets for commercial aircraft. There is extensive collaboration between the aircraft manufacturers, often termed OEMs, and their potential customers. These are usually airlines, but may include other operators. The ultimate customers, however, are passengers, and their requirements are determined by surveys.

The marketing specialists use these inputs and other data to study global economic and population trends. These can be used to predict patterns of passenger and cargo growth or contraction. There is usually a close correlation between GDP and Revenue Passenger Kilometers. These trends are analyzed at global and regional levels and typically extrapolated over a 20-year period. This gives estimates of the total number of aircraft available seat miles (ASMs) required. Studies are then performed to determine route structures, aircraft flight frequencies, and therefore the required individual aircraft passenger-range capabilities.

Figure 3 [4] shows an extract from Boeing's market predictions for the period between 2016 and 2036. It estimates the total market for each world region, together

FIGURE 3 Commercial aircraft market predictions for 2016-2036 [4].

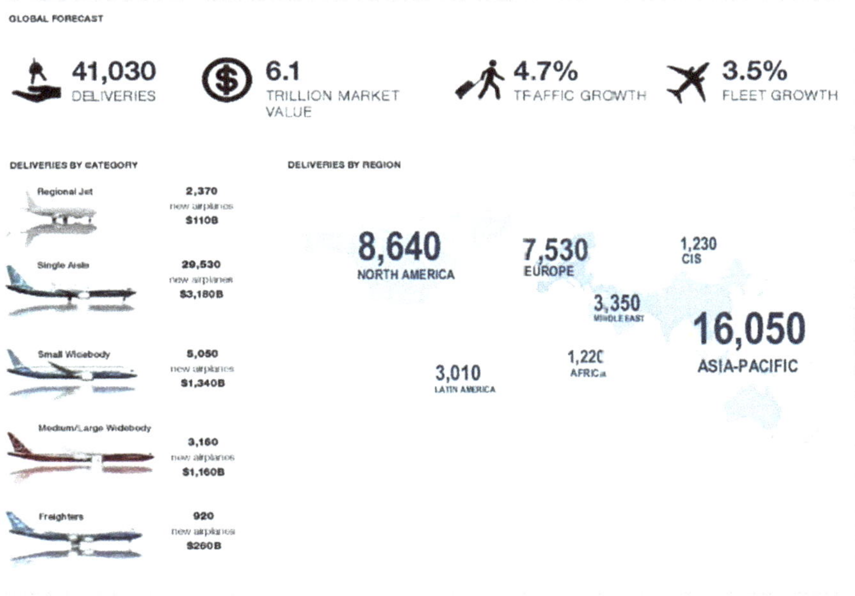

with globally required aircraft numbers of different classes of aircraft. These extend from short-range regional jets carrying some 70-130 passengers to long-range aircraft with between 400 and 600 passengers, and to cargo aircraft. Such statistics allow for the fact that there will be many operational aircraft flying at the start of the prediction period. Estimates are made of these aircraft retirement dates. They are progressively removed from the numbers of aircraft needed to arrive at the net figure of required aircraft shown in Figure 3.

The next stage is the difficult one for prediction of the market share for each OEM, in the category of aircraft to be designed. Previous experience may be a useful guide. In recent years a duopoly has emerged between Boeing and Airbus, with approximately equal shares of sales. This is a simplification, as neither companies offer smaller regional jets, and there has been determined development of larger aircraft from countries including China and Russia. Market share is also affected by the attractiveness of the new aircraft in terms of cost and comfort, which are driven by several factors, including cost and passenger acceptance. A similar duopoly has developed in the regional aircraft market, between Bombardier and Embraer, with again challenges from new market entrants.

Safety and Environmental Requirements: Aviation safety is the most important requirement for aircraft operations. It covers all aspects of operations, including design, manufacturing, maintenance, and flight. Many lessons have been learned from past accidents and incidents. Most have led to changes in airworthiness requirements, which control the design part of the aerospace business. To obtain a Type Certificate, designs must be shown to be compliant with these requirements. This must be demonstrated by extensive (and expensive!) calculations, simulations, or tests, as appropriate. Aircraft **cannot** be operated commercially without a Type Certificate, thus non-compliance will lead to financial disaster for an OEM.

Global warming, noise, and local air pollution are becoming increasingly important to aircraft design and operations. Means to improve the designs of less-polluting aircraft will be discussed in the new technology section of this chapter below. Wider aspects of certification, safety, and environmental impacts are contained in Chapters 6, 9, and 10 of this book.

Aircraft Top Level Requirements (TLRs): Having determined the above requirements, it is then necessary to determine what are sometimes termed the "Cardinal Points" of the requirements. They will be driven by the performance of existing competitive designs and specific airline requirements, such as fuel burn and Direct Operating Costs (DOC) reductions. 15% is a typical airline requirement for the latter, relative to existing aircraft. Other TLRs include payload range, speed, and field performance and are the main inputs into the conceptual design process.

Conceptual Design: Determination of the Initial Layout

Figure 4 shows the initial layout of a short-range airliner, powered by fuel-efficient unducted fan engines. Some of its and other airliner features will be discussed below.

Most airliners operate in the high-subsonic regime and usually employ **aft-swept wings**, which delay aerodynamic drag rise to acceptable levels. Modern aircraft use

FIGURE 4 Cranfield 150-seat airliner conceptual design.

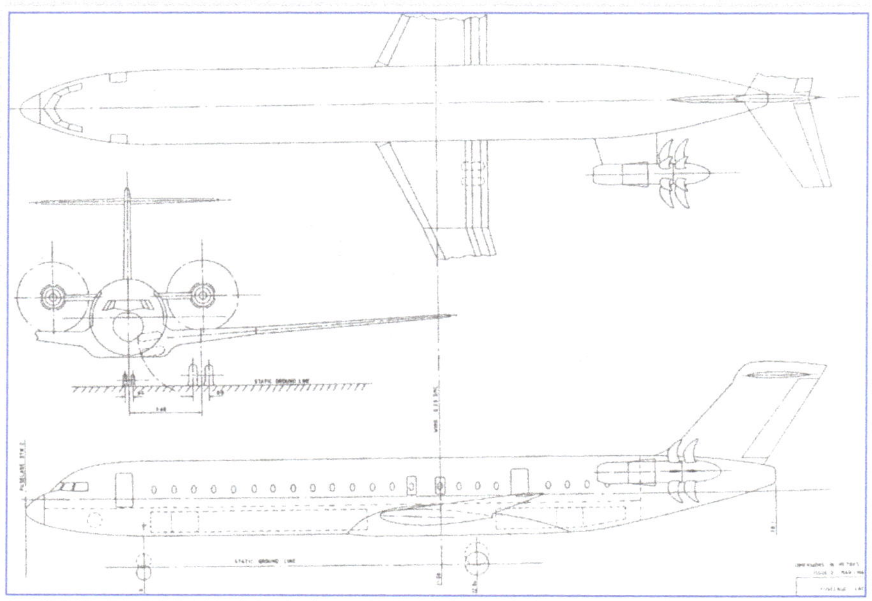

secret airfoil sections, as they are vital to producing superior aerodynamic and therefore economic performance. Initial design determines the wings' geometric taper and aspect ratio (slenderness). These are refined using computational and experimental methods.

The **fuselage** layout is dictated largely by the requirements of passenger safety, space, and comfort, together with adequate baggage and cargo capacity. The designer must obtain the maximum flexibility of layout; thus, the cabin should be unobstructed and parallel if possible. This has the additional merit that it enables extension or reduction of the cabin length to be cheaply made, therefore easing the production of a "family" of aircraft (see later). The cross section of the fuselage is dependent upon the total passenger capacity required, as there is a well-established range of length-to-width ratio for a given type of aircraft. Although there may occasionally be justification in considering a two-deck arrangement for the passengers, it is more usual for passengers to be carried on a single deck with the under-floor volume utilized for freight. Most cabins must be pressurized, and the cross-section shape is therefore best made up of circular arcs. **Seats** vary according to the standard of accommodation. On long-haul routes the economy-class (or Coach) seating is usually placed at about 0.81m (32″) pitch, but for shorter flights this is reduced to as low as 0.7m (28″). **Doors** must be provided for access to the aircraft services, such as galley and toilets, and may also be used as **emergency exits**. The types and sizes are defined in the relevant airworthiness requirements and depend upon the number of passengers carried. They may include the doors and incorporate windows, and should be positioned to help evacuation, for example, over the wing. Adequate toilet and galley facilities are necessary. At least one toilet for every 50 passengers is required on short flights, and more on longer flights.

An early design decision to be made is **the vertical and horizontal location of the wing**, relative to the fuselage. Structurally, the only loads which must be passed from the wing to the fuselage are the wing shear forces produced by lift, drag and side loads, and asymmetric bending moments and torques. It is desirable to have an unbroken wing box, as this saves structural complexity and weight.

Aerodynamically, a mid-wing position is efficient, in that it will have lower interference drag, and this is particularly advantageous for a supersonic aircraft. Mid-wings, however, lead to severe structural or cabin layout restrictions and are not used on current airliners. A wing mounted high on the fuselage may have a favorable critical Mach number due to the absence of fuselage area interference on the top surface. Interference drag may be reduced by careful fairing between either high or low wing, and their aerodynamic efficiency is comparable. A low-mounted wing gives better flotation after ditching in water, as demonstrated by "The Miracle on the Hudson" [5].

From a clearance point of view, the high wing is beneficial when wing-mounted **propeller engines or very high bypass ratio engines** are used. This also ensures that wing fuel is well away from the ground in the event of a crash, but the fuselage must react to all the crash loads. A high wing provides a good view from cabin windows and good all-round access. Attachment of a high wing across the top of the fuselage may restrict cabin head room and may introduce landing gear layout problems.

When a tailplane (or horizontal stabilizer) is used, its efficiency is dependent upon its **vertical location** relative to the wing's downwash. On conventional layouts there is no difficulty in locating the tailplane roughly in the wing plane. In practice, it is often slightly higher as this improves rear fuselage and tail ground clearance. When an aircraft has **aft fuselage-mounted engines** (as in Figure 4), the need for engine exhaust clearance means that this can only be done by using a high-mounted or cruciform tailplane. In this case great care must be taken to ensure that dynamic pitching of the aircraft to high incidences does not take the tailplane into a severe downwash region. If this occurs a "deep stall" results. Most modern aircraft use what is termed the "Tricycle" Landing gear arrangement, with a nose gear used for support and steering. Main landing gears on low-wing aircraft are usually attached to their inner wings, but the wheels retract into the fuselage. High-wing aircraft usually mount main gears on the lower fuselage, often in blister fairings.

Working Together to Design the Aircraft

The development of any new aircraft is a challenging, expensive, and time-consuming task. This is particularly true for high-subsonic airliners. The Airbus A380 and Boeing 787 design and development processes required several years of work by thousands of engineers scattered round the world, costing billions of dollars. In many cases there is direct or indirect financial support from national governments, but there is a need for global participation. The OEMs need to develop teams of partners to deliver the aircraft. This is not just to share financial or technical risks but also the need for technical resources and competences, not available to OEMs. In the early days of airliner design, the whole of the aircraft could be produced by individual companies, such as De Havilland of the United Kingdom. It produced airframes, engines, landing gears, and propellers, although many system components were produced by specialist suppliers.

Modern aircraft are clearly much more complex, and specialized companies have evolved to produce engines, landing gears, avionics, and airframe system components. Even aircraft structural assemblies are produced by the OEM's partner or sub-contract companies. These are often based in different parts of the world and require considerable collaboration and integration. Funding issues can also require complex negotiations to determine equitable work share and clear agreements for the required program deliverables.

The airlines are vital contributors to the derivation of the aircraft requirements. There also needs to be close collaboration between customer airlines and OEMs and suppliers at all stages of design and production. A typical example is their collaboration on the development of the aircraft's maintenance activities by means of the MSG-3 process, described elsewhere in this book.

Integration of all partners is enabled by using digital product definitions, using shared Computer-Aided Design and Computer-Aided Manufacturing (CAD/CAM) systems.

Physical co-location of designers is ideal, but not often possible. Modern communication methods help, but allowances must be made for time-zone and cultural differences around the world. Face-to-face contact is vital to building confidence and teamwork, which necessitates considerable travelling between partner organizations.

Design to Meet or Exceed Passenger Expectations

Design for Safety

Airworthiness requirements continue to evolve, as the result of incidents and accidents, and drive safety improvements. One recent example was the introduction of "terrorist-resistant" flight deck doors, as the result of the 9/11 attacks. Measures are being investigated to provide aircraft position information, to prevent the situation that resulted from the loss of flight MH370. Avionic system requirements ensure reduction to the risks of collision with other aircraft, or the ground.

The safety and environmental requirements are crucial, and the design to meet, or exceed, them is the aircraft designer's priority. Initial design work is confirmed by extensive calculations, backed up by computer simulations. Modelling is then verified by computational fluid dynamics (CFD) and wind-tunnel testing for aerodynamic performance.

Parallel work for structures is performed by means of finite element analyses (FEA) and then coupon, sub-assembly, and then whole-aircraft structural testing.

Avionic and system components are modelled and coded as part of preliminary and detailed design, followed by individual component environmental testing. Extensive use is then made of "Iron Bird" test rigs, in which components are integrated into ground-based test rigs, in the correct geographical relationship to other components. These are then powered up and their performance tested. This process is continued after the installation of engines and systems into the prototype aircraft, and then during flight test.

Results from these tests are fed back to designers by an iterative process, so that the design may be modified. The airworthiness authorities closely monitor the whole design and development process and, if satisfied, issue a Type Certificate, which then allows the design to proceed to manufacture, sale, and service.

This process is extremely demanding of time, resources, and expertise. It provides a large barrier to new entrants into aircraft manufacture. There are recent examples of otherwise competent designs being unable to obtain Type Certificates and the programs failing.

Design to Reduce Passenger Costs, While Delivering Profits to Manufacturers and Airlines

DOC Elements: Figure 5 shows typical DOC percentages of a single-aisle short-haul airliner.

The most volatile value is that of fuel costs, which have varied enormously in recent years, linked to crude oil values varying between $130 and $30 per barrel. These costs do not include future environmental taxes. Design methods for DOC reductions are discussed below.

FIGURE 5 Direct operating cost.

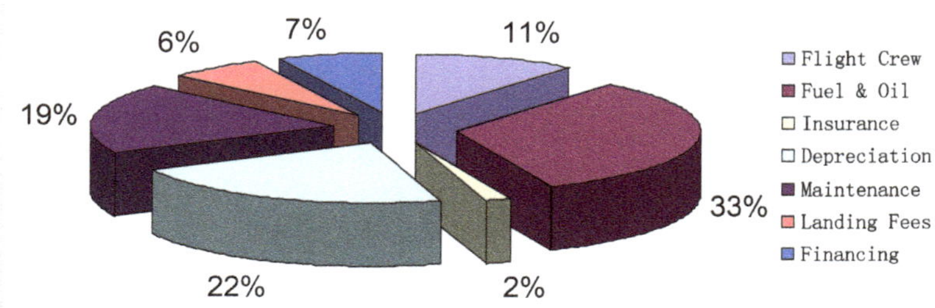

Reducing Fuel and Oil Costs and Commensurate CO_2 Reductions: Modern aircraft design technology and new aircraft and engine configurations can bring significant fuel usage, and therefore pollution, reductions. Development areas include improved **aerodynamic shapes**, including the elusive "laminar flow" and increased wing aspect ratios, to reduce induced drag. Reduced aircraft mass will reduce the required wing sizes with consequent drag reduction. Further mass-reduction technologies include lightweight composite and metallic structures, advanced systems, and avionics. The most significant effect, however, results from engine improvements. Recent engines have greatly increased bypass ratios, temperatures, and pressure ratios. Figure 11 shows modern "Open Rotor" engines fitted to a possible future environmentally friendly airliner. Such propulsion systems have several designations, including "Prop-fans" and "Ultra High Bypass Ratio."

The business case for new engines is highly dependent on fuel prices. Higher bypass engines are heavier and more expensive than their predecessors. For example, the Airbus A320 New Engine Option (neo) costed $108.4M in 2017, while the A320 Conventional Engine Option (ceo) costed $99M. With low fuel prices, it's better to use the cheaper-priced ceo aircraft or even refurbish Mojave Desert-stored surplus aircraft.

The neo has a heavier Operating Empty Mass, but the same All Up Mass. With a fuel burn some 15% better, it however has significantly better range, increasing its operational attractiveness. Increased fuel prices make it even more attractive, especially as it produces significantly less global warming.

Novel configurations offer even more performance benefits, as discussed later.

Reducing Depreciation Costs: Airliners are very expensive to buy, so two approaches have been developed to pay for them. Airlines **depreciate** the **acquisition costs** over the life of the aircraft, say for some 20 years. This accounts for the depreciation element of the DOC costs. An alternative is to rent the aircraft from a leasing company. The **acquisition cost** of an aircraft includes **development** and **manufacturing** costs (see below). Current **list prices** are shown in Figure 6, but complex discount deals are usually the norm, and therefore acquisition costs are often less than list prices. It can be seen from the figure that there seems to be a direct correlation between the price and the Operating Empty Weight of the aircraft. This is an indicator of the complexity and efficiency of the aircraft.

Development costs of the aircraft include design, testing, production of prototype and structural test aircraft, and associated flight tests. These costs are independent of

FIGURE 6 Airliner list prices, 2016 [5].

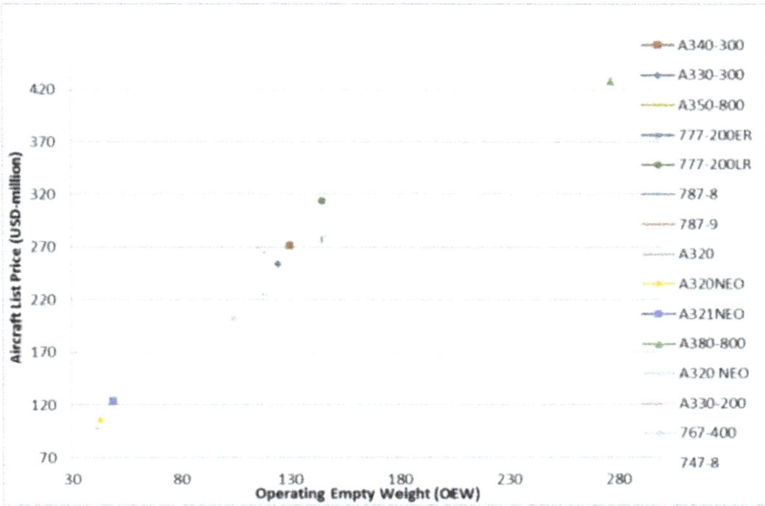

the number of production aircraft delivered, and their costs must be spread over the income from aircraft sales. Recouping development costs is achieved when the program reached the break-even point. This varies with aircraft types and OEM, but is of the order of 400 aircraft.

Manufacturing costs cover the costs of materials, labor, engines, and avionic and airframe systems, together with testing and inspection. There has been recent significant progress in reducing such costs. The main factors are better design for manufacture and assembly, increased production runs, and optimization of the balance between automation and labor costs in manufacture and assembly. New techniques have been developed for composite material component manufacture and assembly, which include filament winding, tape laying machines, and resin infusion. Modern CAD/CAM facilities improve accuracy and can lead to "Jigless" manufacture, with significant cost and time savings, as do Just-in-Time (JIT) approaches. The JIT assembly process carefully controls the timing of the supply chain delivery of components, leaving the most expensive items, such as engines until the last moment. There have, however, been occurrences of JIT, when components are delivered too late! An emerging technology is that of material deposition, sometimes termed additive manufacture. It is a 3D printing process for the delivery of complex-shaped components.

Reducing insurance costs are based on risk and are linked to the acquisition cost, airline reputation, and route structure. Cost savings will directly follow acquisition cost reductions and safe operations.

Reducing landing fees relate to the aircraft's mass and physical dimensions; thus, a lighter aircraft is cheaper to buy, uses less fuel, and costs less to land. There are standard airfield categories in terms of aircraft wingspan and length, which can impose severe constraints on designs.

Reducing Maintenance Costs: Figure 7 shows the main aspects of design for good maintainability, which are used as design guidelines. The most significant

FIGURE 7 Design for maintainability requirements and CAD manikin.

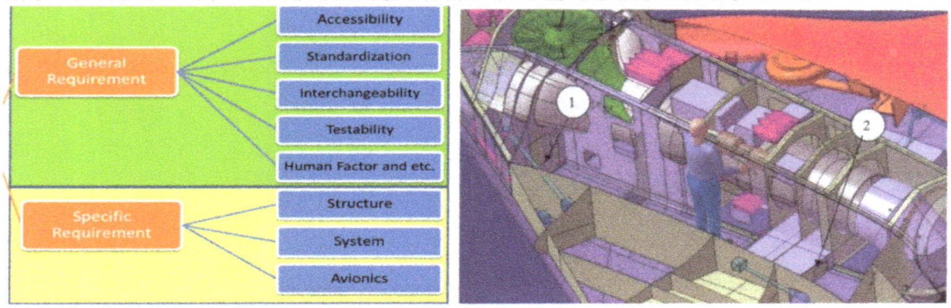

requirement is to produce good access for diagnosis, rectification, or replacement of components. CAD digital mock-ups have largely replaced full-scale physical mock-ups. They can be used much earlier in the design process, can highlight clashes, and are infinitely scalable. Figure 7 also shows a CAD manikin, which can be used to check for access and maintenance actions.

Maintainability design also has inputs into the types of maintenance required and their frequencies, as discussed in another chapter of this book. Work is also required for the design of assembly and heavy maintenance staging and specialized Ground Support Equipment (GSE). Modern technology, such as IVHM, can significantly improve fault diagnosis for components and software, thus reducing downtime and maintenance costs. This follows decades of improved engine and aircraft structure health monitoring.

Reducing Crew Costs: Developments in avionics and flight control systems design have resulted in safer and more efficient flight decks, such as that of a Boeing 777 as shown in Figure 8. Cross-qualification of flight crews across families of aircraft are mentioned below, which eases crew rostering and reduces training costs.

There have been improvements in the control and navigation of aircraft, leading to the gradual reduction in the number of required flight crew members, with consequent cost savings. It is possible to consider further reduction of these costs by the introduction of single-pilot or no-pilot operations. This clearly has airworthiness and passenger acceptance issues! More simulator training.

FIGURE 8 One of the authors sitting at the flight deck of a Boeing 777.

Developing "families" of aircraft positively affect manufacturing, development, crew training, maintenance, and certification costs. Most airliner OEMs produce "families" of aircraft, which cover wide spectra of payload-range options. The Boeing single-aisle family does this by using identical fuselage cross sections in its 707, 720, 727, 737, and 757 aircraft. This considerably reduces manufacturing costs and gives improved commonality in service. A similar approach is followed by the Airbus A318, A319, A320, and A321 family. The "family" is produced by stretching or shrinking the parallel section of the fuselage by inserting "plugs" of fuselage ahead and behind the wing. This increases the capacity significantly for relatively little extra cost. Such an approach also maximizes the commonality of avionics and airframe system layout and components. This reduces manufacturing, spare part provision, and maintenance costs. This also reduces aircrew and maintenance engineering training costs, as crews can be cross-qualified across a family of aircraft. This approach is taken further by re-engining older aircraft types, as exhibited by the Boeing 737MAX, Boeing 777, and Airbus A320 and A330neo families. This process is not always easy when incorporating larger-diameter engines on low-wing aircraft, as it may result in costly changes to landing gears. Airbus followed an additional family approach by using a common wing design for the twin-engine A330 and four-engine A340.

While fuselage external dimensions are fixed, there is considerable scope to vary the internal arrangements. Careful design may introduce extra seat rows, or an extra seat per seat row. This may affect passenger comfort, but will increase passenger capacity and profits for the airline.

Further profitable family members follow from conversion to other roles, such as main-deck cargo, corporate jets, or firefighting aircraft. Military derivatives are also possible, such as maritime reconnaissance, electronic warfare, or flight-refueling aircraft.

Design for Comfort and Communications

The choice of fuselage cross section has been mentioned above, as it is a crucial factor in passenger comfort and cost. Passengers, naturally, want as much space as possible, but there are limits on fuselage size in terms of aerodynamic drag and structural mass. The critical case is that of the economy-class cabin. The Boeing 737 cross section is the baseline for single-aisle cabins, the width of which has been slightly exceeded by the later Airbus A320 family. Figure 10 shows a more recent Cranfield University project aircraft, which goes further in space. Recent proposals have included staggered three-seat row seats, with the unpopular center seat being wider and set back, relative to the side seats. The other major space comfort factor is that of seat pitch. Economy-class pitches remain at some 32 inches, but premium class seat pitches continue to increase, as do prices! First-class passengers have many facilities, including an onboard shower provided by certain operators of the Airbus A380 (also shown in **Figure 9**).

Further issues of comfort include design for reduced internal noise, lower cabin altitudes, higher humidity, and better furnishings. In-flight entertainment continues to

FIGURE 9 Cabin interiors—Cranfield "Flying Crane" project and Airbus A380 shower.

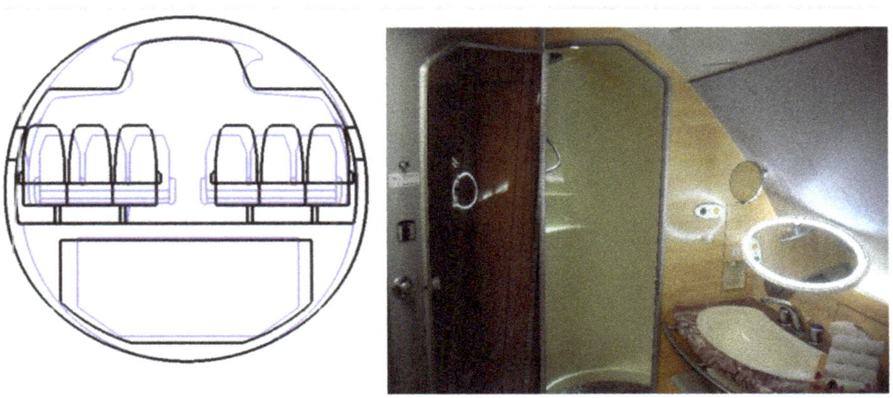

FIGURE 10 Development of airliner cruise speeds.

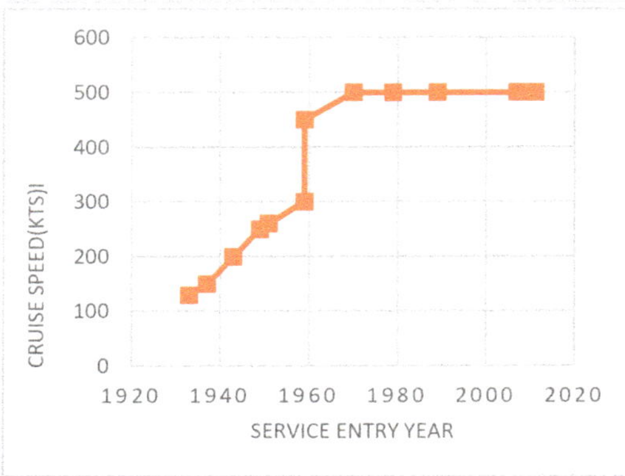

be improved, as has digital connectivity, enabling passengers to continue working or enjoying entertainment on the aircraft.

Design for Speed and Flight Frequencies

Speed has always been the primary motivation for passengers, since the dawn of flight. **Figure 10** shows progress since the start of airline operations. Propeller-driven aircraft gradually increased speed, but there was a step change with the advent of jet-powered aircraft. Except for the Concorde supersonic airliner, cruise speeds have now reached a plateau.

Cruise speed, however, is only part of the journey time. A salutary fact is that journey time between central London and central Paris today takes about 4 hours, the same time it took in 1919, flying in a converted World War I biplane bomber! This is an example of the threat to air transport provided by high-speed rail. Other chapters of this book discuss means of reducing journey times through airports, but the designer can help! Turnaround times at airports may be significantly improved by fuselage layout, design for maintainability, and fuel system design. Airport capacity is a growing constraint on passenger travel. This may be alleviated by using larger-capacity aircraft, but does not help to increase flight frequencies that may give more choice of when passengers may fly. Extra runways can help, but they are controversial.

Design for Environmental Responsibility

The impact of aviation on the environment has been a major concern for at least two decades. Negative public opinion is growing rapidly and significant steps are required to alleviate aircraft emissions in a growing air transport market.

A significant initiative was taken in Europe in 2001, with the publication of quantified targets. These were summarized in [6] as:

"Europe…has its Advisory Council for Aeronautical Research in Europe (ACARE), which, in 2001 committed the industry to developing technologies that would cut CO_2 by 50% per passenger kilometre, cut NO_X by 80% and halve perceived noise levels by 2020, relative to 2000 standards." This process has been extended to the Flightpath 2050 and other requirements, as discussed in a later chapter. Some means of making progress toward meeting noise and carbon targets are shown below.

Introduction of Possible New Technologies and Configurations

Current airliners conform to a relatively limited number of configurations, typified by "wing and tube" designs. These have limited potential to meet future challenging

FIGURE 11 Some Cranfield University advanced airliner technologies and configurations.

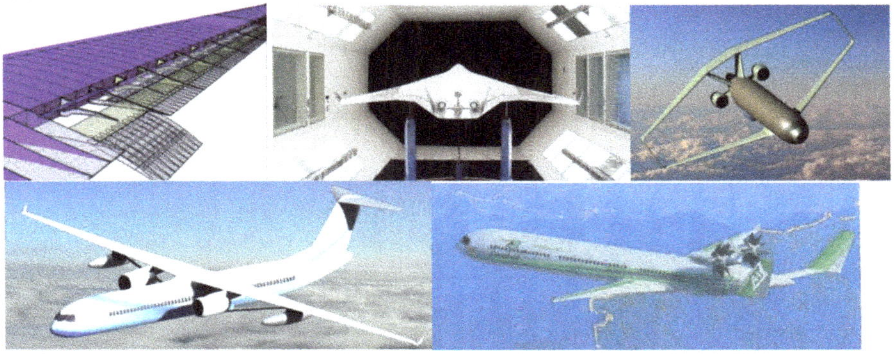

environmental requirements. **Figure 11** shows some possible ways forward. Conventional configurations are being improved by the development of technologies such as aerodynamic laminar flow, variable-camber wings, and ultra-high bypass engines, which are shown on the bottom right of the figure. The top left image is a proposed variable-camber flap system, which optimizes the use of flaps throughout the whole of flight, including cruise. The bottom left project utilizes an extremely high aspect ratio wing to reduce induced drag and uses liquefied natural gas as an alternative fuel. The more radical shapes shown top-right are blended-wing-body and box-wing configurations. These offer more economic and environmental benefits, but at considerable project risk. However, risks must be taken to meet the challenges, and aircraft designers can help.

References

1. Anon, *CLARUS Concept of Operations*, (Washington: Federal Highways Administration, 2005).
2. Haberland, C. Thorbeck, J., and Fenske, W.A., "Computer Augmented Procedure for Commercial Aircraft Preliminary Design and Optimization," *14th Conference of the International Congress of Aeronautical Sciences, Paper ICAS*, Jerusalem, Israel, 1984.
3. Fielding, J.P., *Introduction to Aircraft Design*, 2nd edn. (UK: Cambridge University Press, 2017).
4. Anon, *Current Market Outlook - 2017*, (Washington: Boeing Commercial Airplane Company, 2017).
5. Chesley, B., "Sullenberger 111 with Jeffrey Zaslow, Highest Duty-My Search for What Really Matters," *Harper*, 2009.
6. Turner, A, "Changed Climate-Special Report," *Flight International Magazine*, June 19-25, 2007.

CHAPTER 4

The Planemakers: Manufacturing the Products

Richard Gardner
Farnborough Air Sciences Trust

I suppose we shall soon travel by air-vessels: make air instead of sea voyages; and at length find our way to the moon….

Lord Byron 1882

Turning the dream of aviation into a practical reality was initially restricted by the lack of knowledge on such key issues as material and structural strength, methods of control, and sufficient power to provide initial lift off the ground and sustained propulsion through the air. While most of the basic aerodynamic requirements for heavier-than-air powered flight were understood in the late 19th century, the industrial revolution had already created tools and machines capable of mass-producing complex mechanical devices. Applying this new technology to aviation was an obvious next step. The main aviation challenge was to combine vehicle strength with engines that had a power-to-weight ratio that would allow a flying machine to carry a pilot as well as its own all-up weight. This was just the dawn of aviation and the pioneers were competitively struggling to fly machines that were safe and which could be developed further. All their early designs were hand-built, mostly involving fabric-covered wooden frames. Lightweight metals, including aluminum, appeared early in the first decade of the 20th century and were soon applied to the design of new engines and airplane fittings.

Powering Up a New Industry

The first experimental aero engines featured steam power, but were too heavy and bulky. The need for weight reduction was a major motivation, and as the successful Wright brothers realized, the internal combustion engine was seen as a key enabling factor. Hence, the first obstacle to overcome before a powered airplane could be put into production was therefore identified early on. The availability of lightweight engines opened the door to suitable engine designs that would be relatively simple to manufacture, operate, maintain, and repair. The trickle of new airplane designs became a flood within a decade, as more powerful engines appeared, and a host of new manufacturers emerged, anxious to share in what was seen as an exciting and potentially profitable new industry. Aviation was clearly destined to be the next big game-changer. Many companies had started businesses building motor cars and goods vehicles. Others had been engaged in marine production, while some launched aviation activities as an off-shoot from established engineering work. All these enterprises brought great skills with them, including manufacturing complex products, such as gearboxes, fuel pumps, radiators, and a number of cast or machined fittings of every shape and size. Suppliers of cables, fabric coverings, paint and dopes, as well as seats, flying clothing, goggles, helmets, metal panels, wheels, brakes, and springs all prospered as more and more specialized products were developed to meet the needs of the new airplane constructors.

Specialization Grows

Aero component designs were prepared in the drawing offices of the new companies using well-tried engineering principles. Alternative arrangements for important features were often developed as the process progressed. In the early years, there was little experience for the industry's workers to rely on beyond what they had learned from working on other engineering products or the study of technical papers and articles. However, it was not long before the complexities of aeronautics demanded more specialists becoming part of the workforce. Aviation was a new science, so at first there was no recognized industry standard of training, and new apprentices had to work alongside those who were still pioneering methods themselves. Special knowledge in physics, chemistry, metallurgy, and engines became essential as aviation developed. The establishment of formal standards and specifications to be used in aeronautical construction was first devised and made official at the Royal Aircraft Factory at Farnborough, in England in 1914 [1]. Initially, these standards and specifications were needed to devise rational methods of stress calculations that were introduced into the technique of design and strength testing of complete structures. This introduction was an essential building block for the future. Building airplanes that were stable in flight and mechanically reliable was a big step, but by the end of the first decade of powered flight, this was being achieved by designs that were soon to be built in sufficient numbers to be sent to war World War I had been the most significant milestone in the development of aviation as we now know it, for it had resulted in the mass production of thousands of airplanes in new factories in Europe and the United States. Total airplane construction during this period was: France 67,987; United Kingdom 58,144; Germany 48,537; Italy 20,000; United States 15,000 [2].

At that time, the factory production lines had been set up over a relatively short timescale, with major sub-assemblies and components, including engines, being produced elsewhere, coming together in final assembly where they were fully fitted out and prepared for their initial clearance test flight. As more standardization of aircraft requirements and specifications became adopted, it was easier for sub-contracting companies to take on aeronautical manufacturing jobs to assist in meeting the demands that war had generated. As a result, established engineering, automobile, and even furniture companies became part of the expanding aviation supply chain. Airplane factories were very large, and most also had their own airfield facilities to deliver completed machines. They also had to have good road, railway, or navigable waterways on hand to allow the constant stream of raw materials and finished components to be handled. These factories required adequate local power to be available for electrical and gas supplies, and a source of workers to populate the factory. As this was all new industry and existing workshops and final assembly halls were already working to capacity on war work, the arrival of aviation manufacturing resulted in the large-scale recruiting of women to help compensate for the shortage of male workers. This in itself was a revolutionary change in work practices at the time, but the women soon became very expert in a host of special skills, including machining work on industrial lathes and drills and assembling components. At the time, some of the most well-known airplane manufacturers included Avro, Bleriot, Sopwith, Curtiss, Bristol, De Havilland, Fokker, Morane-Saulnier, Nieuport, Short Brothers, SPAD, Handley Page, and Vickers. Early production airplane structures had been constructed mostly of wood and canvas, and a jig assembly maintained consistency in build standards. As performance of the airplanes improved gradually, the use of metallic parts increased, bringing more strength while keeping structural weight down. Metal structural frames, as well as shaped cowlings, became more widely adopted, and this again repeated much of the advancement in material use that had been imported from the automotive sector.

Birth of a New Industry

When WWI ended in 1918, after becoming the first conflict in history in which aviation played a pivotal role, the aeronautical industry found itself poised on the cliff edge of abandonment. Thousands of military airplanes were no longer needed and neither was the vast production capacity provided by hundreds of purpose-built factories and the massive component supply chain. The factories were built to deliver as many airplanes as possible, but suddenly contracts were cancelled, and the factories faced wholesale closure. There was no commercial airline industry to take up the slack, and this was a high-risk area for investors as there were so many unknowns, despite the obvious potential of air transport. The first airline was set up in the United States on December 17, 1913, and on January 1, 1914, the Tampa Airboat Line began a scheduled service between St Petersburg and Tampa. The passenger service lasted just three months and interest switched to the carriage of mail by air. A few mail-carrying services, which started in May 1918 and were flown by the U.S. Army, were followed in 1919 with a passenger flying boat service between New York and Atlantic City. In Europe, similar early mail and air transport flights had opened between London and Paris in 1919 [3]. These flights were made by small airplanes, mostly converted from former military light bombers, such as the DH4. This set a pattern in the early 1920s as new operators sought to adapt the largest wartime bombers for a new commercial use. The cost of developing new designs was a mix of private and public funding. The largest airplane manufacturers raised their own investment funding as they adopted more modern methods and developments, but this private enterprise was against a backdrop of continued state support, especially as aviation was regarded as a matter of national pride and international prestige. The emerging commercial airplanes were also able to benefit from new engine and other developments that had been assisted by military research and development (R&D) investment. The Vickers Vimy made the first successful non-stop transatlantic crossing in 1919, demonstrating that long-range air travel was now a possibility. With few new demands for military aircraft, the manufacturers looked toward the potential of commercial air travel to create a whole new market with designs capable of carrying an economical payload of passengers and cargo. Both flying boats and land airplanes would be needed, with safety and reliability becoming the most important requirements to establish international air routes that a travelling public could support with confidence. The post-war years from 1929 were dominated by trade depression, large-scale unemployment, and industrial contraction, which inevitably led to further pressure on funding for investment in new aviation technology. This period of retrenchment in the industry led to more regulation. In the United States, the Kelly Act had led to the 1926 Air Commerce Act which promoted expansion of the U.S. Air Mail service. Also, a nationwide network of lighting beacons greatly assisted the development of long-distance air flights, which the manufacturers were keen to exploit. The early aviation companies in the United States combined air services with manufacturing, which gave them a greater capital base for investment. The Boeing Air Transport Company, formed in 1927, merged with the United Aircraft and Transportation Company which became United Air Lines (Figure 1). Boeing built 24 new Model 40A mail planes, but mail carrying and passenger services were not yet profitable. Ford entered the commercial market with its tri-motor airliner, which proved to be a popular machine, opening up many transcontinental air routes. In June 1934, Congress passed the Air Mail Act which separated manufacturers from airlines [4].

FIGURE 1 The original Boeing factory has been preserved at Seattle.

Understandably, manufacturers initially tended to base their commercial aircraft on proven design parameters. This led to designs having at least two engines, in case of failure of one, and being of a wire-braced biplane configuration for strength. Building such airplanes posed no great technical barriers, but they would need to be larger and would not have any speed advantage over older designs, typically having a top cruising speed of little over 120 mph. While the manufacture of airplanes in the 1920s can be considered a non-innovative, evolutionary era, by the end of the decade, air travel was starting to acquire an image of excitement and attracting increasing public interest. Initially, only the rich could afford to sample its delights, but manufacturers were looking to stepping up their game. The transition arrived in the 1930s. More attention was given to the use of different steels and light alloys even though much fuselage and wing surface covering was still made of fabric. As the materials changed, the skill requirements of the workforce also had to change. Mass production of airplanes, made increasingly of metal, enabled designers to break free from many traditional constraints. Metal-forming techniques allowed more streamlined aerodynamic shapes to be incorporated, which reduced drag, increased speed, and improved fuel efficiency. Strengthened metal wing boxes, spars, and stressed skin wings brought forward more use of monoplane designs, which further reduced drag and increased speed. A modern breakthrough in design, combining most of the new techniques, was the Boeing Model 247 which was sold to United Air Lines in 1933. This aircraft was powered by two 550 hp Pratt & Whitney Wasp radial engines and was capable of carrying ten passengers at 160 mph. As the main manufacturers, including Lockheed, Dornier, Fokker, Vickers, Handley Page, Martin, Junkers, De Havilland, Douglas, Wibault, Shorts, and Sikorsky, announced their latest civil airplanes, new opportunities were opened up for suppliers of engines, instruments, and associated equipment. The aviation industry was becoming interdependent and global and increasingly competitive. A resurgence in nationalism in Europe had resulted in significant re-armament programs which promoted the development of new aircraft designs that could be built in military versions, such as bombers or reconnaissance aircraft, or in transport either as troop carriers or commercial passenger airliners. One of the most popular versions was the Junkers Ju-52 "Iron Annie" monoplane featuring three engines and a corrugated metal fuselage and wings (Figure 2). This design popularity was particularly good news for manufacturers in the early and mid-1930s, who could bring to the market much more modern metal aircraft that could fly long distances at higher speeds than previously. These modern aircraft were produced in large numbers from well-organized assembly lines, planned with future war production levels in mind, where the delivery rates could be increased rapidly with provision for handling inward component supplies, with generous storage areas as a part of the factory facilities.

FIGURE 2 A line-up of typical mid-1930s civil aircraft at Croydon, England, during the crossover period between biplane and monoplane designs with all-metal replacing wood, canvas, and metal structures.

In the United States, air travel was about to become the dominant long- and medium-distance preferred mode of transport, in competition with rail travel. Monoplanes, such as the Douglas DC-1 and DC-2 and the Lockheed Electra, represented the most modern designs available with all-metal structures, powerful engines, and retractable landing gear. Large manufacturing facilities would soon be expanded further to cater for previously unimaginable levels of

FIGURE 3 A typical late 1940s-1950s production line of piston-engine airliners, featuring Douglas DC-7s.

production output during WWII. As this growth in industrial capacity continued, the aviation industry in the United States was very active building large numbers of transport aircraft covering every size and range category. By 1939, around 80% of air passengers in the United States were flying aboard DC-3s. But, in 1945, unlike 1918, the many successful transport aircraft in production had an immediate appeal for companies and state airlines wishing to resume civil air operations. This demand kept production lines flowing for piston-engine civil transports well into the 1950s, by which time the new jet-age era transformed the civil aviation market (Figure 3).

FIGURE 4 The turboprop Vickers Viscount (left) and jet DH Comet were the first civil gas turbine-powered airliners.

Building for the Jet Age

The new generation of pure-jet and turboprop airliners, including the DH Comet, the Boeing 707, the Douglas DC-8, and the Vickers Viscount, brought smooth, above-the-weather passenger comfort on all air routes, and the production lines kept on rolling as air fleets expanded (Figure 4). The manufacturing process had changed remarkably little in the transition to the jets, and although the engines were easier to build and operate, the airframes had become more complex with pressurized cabins and more avionics and other electrical equipment, such as radar. Final assembly was still very labor intensive, where completed airframes were fitted out with all the electrical and hydraulic systems added. Another manufacturing development that became widespread in the 1950s and 1960s was licensed production.

This approved process enabled total production levels to be increased without the OEMs having to invest in new assembly facilities, allowing the licensed manufacturer to embark on further sales campaigns to expand the market, especially on a regional basis. The OEMs gained through royalty payments and also in the manufacture of further spares and replacement components throughout the life of the aircraft.

During the wartime ramp-ups in production output, many final assembly lines incorporated primary structures that had been already fitted out with many systems that were built off-site and then transported to the final assembly line for bolting together and completion. Modular assembly had been slower to apply to civil programs post-war, but it made a comeback as orders increased and more sub-contracting led to alternative sources of supply for major structural components, such as wings and fuselage sections. In the United Kingdom, where factories had been priority targets for German bombers during the war, the government supported dispersal policy for aircraft construction and post-war this resulted in a wide distribution of plants all capable of manufacturing major sections, key components, or even complete aircraft. In the 1950s, this dispersal allowed for shared production between separate companies for aircraft required in large numbers and helped to sustain many sites that were threatened by post-war consolidation to reduce overcapacity. As national aviation production became more concentrated in ever larger corporate groups in the United States and Europe, the supply chain nevertheless continued to expand. Early manufacturers had been very self-sufficient, buying in only such items as engines. In some cases even these were produced in-house. By the early 1960s, as the civil air transport market grew, powered by the advances of the jet age, the spread of suppliers extended well beyond national boundaries. Engines were built in the United Kingdom, the United States, Canada, France, and Spain and were being fitted to new civil aircraft made on both sides of the Atlantic and around the world from Sweden to India and Japan to Australia. License-built aircraft were often modified and modernized to meet new local customer requirements, and this involved such changes as alternative engines, new cockpit avionics, and in some cases building cargo-carrying versions with special fittings. A new specialized industry emerged, with engineering companies modifying standard passenger-carrying airliners into cargo carrying transports. This speciality involved increasingly complex work, especially when involving the conversion of aircraft featuring pressurized cabins. Some conversions involved the fitting of complete hinged nose or tail sections, which allowed outsize cargo to be loaded and unloaded easily straight into the cabin hold, and not via a side-fitted door. This conversion activity has subsequently grown into a global business giving retired passenger aircraft a new lease of life as cargo carriers.

Expanding Partnerships

The concept of true international cooperation soon expanded well beyond licensed co-production to full international program cooperation. The global manufacturing era arrived in a big way during the 1960s. This era was seen as a mutually beneficial way of sharing R&D costs between two or more national partners and at the same time enlarging the total domestic "launch" market for all-new airplanes. The most spectacular example was the iconic Anglo-French Concorde supersonic airliner, which first flew in 1969. This program was, and still is, the most ambitious commercial aircraft project ever undertaken from a technological and manufacturing viewpoint, enabling over 100 passengers to travel in normal jetliner comfort at twice the speed of sound at an altitude of 60,000 ft crossing the Atlantic in three hours and 15 minutes. To manufacture such a machine, which flew more than twice as fast and twice as high as a conventional jet, involved years of testing and evaluation. The aircraft manufacturing was split between British and French companies, and there was no duplication of construction sections or components. Different parts of the aircraft were allocated to single sources of supply, but there were two final assembly lines, one in Toulouse, France, and the other at Bristol in the United Kingdom. The engine was a specially designed development of a British high-thrust turbojet, the Olympus. Bristol was responsible for the overall engine integration incorporating a Bristol Siddeley/ Rolls-Royce main core and the afterburning and thrust reversal systems supplied by Snecma of France. The thousands of components made in the United Kingdom and France were all delivered from various separate aircraft factories and supply-chain sources to the two final assembly plants where they were incorporated, tested, and made ready before internal fitting out and painting, and then the completed machine was ready for its pre-delivery test flights. The development and manufacture of the Concorde was undertaken in a pre-digital environment, and while it was the world's most advanced jetliner, its manufacturing process was almost entirely traditional, and extremely labor intensive.

The international cooperation pattern reached a new level with the coming together of the European Airbus consortium in the late 1960s. The group's first product was a 300-seat wide-body airliner and was revolutionary in many ways, not the least being its construction spread across several European countries (with further equipment and the engines produced in the United States). Each industrial partner was responsible for an agreed proportional share in the program and for delivering near complete sections (forward and center fuselage, wings, rear fuselage, and tail) which were flown out to Toulouse, France, for final assembly, systems integration, and test flying. The number of personnel involved in the final assembly stage was very small compared to in a conventional aerospace factory, but the partners could call on the full industrial capability of their own aviation manufacturing facilities to contribute. This product was a highly political project from the start, calling for state investment support from project partner governments, with the long-term aim of creating more European commercial competition to challenge the giant U.S. industry, which had the advantage of large-scale government support for military orders for military transports, which could be developed for the commercial market. As Airbus struggled to win orders for its A300B, the U.S. aviation industry enjoyed a 90% share of the world market for modern commercial aircraft, with many new-generation wide-body products on offer, including Lockheed's L-1011 TriStar, Boeing's 747, and McDonnell Douglas's DC-10. The first Airbus design was built to common drawings in the pre-digital era when original plans were hand prepared in a drawing office by draftsmen and women. But in other respects, the new aircraft introduced many cutting-edge features. Many major structural components were milled from solid alloy billets and the wing aerodynamics were highly advanced giving an excellent take-off/landing and cruise performance with modest runway requirements. It was the first wide-body twin-engine airliner and carried almost twice as many passengers as the then-largest short-haul jetliner, the Boeing 727. It had a 222-inch diameter fuselage cross section which featured generous below-cabin cargo capacity and was powered by highly efficient and quiet big-fan engines. Building it presented European manufacturers with several major challenges. The biggest was how to bring together such large fuselage and wing sub-sections from sites spread all over Europe. Special transport arrangements had to be established at all the main Airbus sites to allow movement of completed sections by a heavy-lift aircraft to Toulouse for final assembly. Ironically, it was a modified Boeing-made transport aircraft that made it all possible. The Guppy had originally been built to carry sections of the Saturn V Apollo moon rocket between locations in the United States, and it was the only cargo airplane large enough at that time capable of carrying the A300B fuselage sections and wings. Airbus gradually acquired a fleet of Guppies, and then followed with modified A300-600s, known as Belugas, to transport fuselages and wings across the world in a huge international manufacturing coordination operation. As Airbus established new final assembly sites in China and the United States, as well as in Europe, its manufacturing support air transport network continued to expand. Today, Boeing also uses outsize cargo aircraft, modified from Boeing 747s, to similarly ferry large sections of its jetliner product line between suppliers in Japan and Italy and its own final assembly sites in Washington state and South Carolina. Other large components destined for Boeing jetliners, such as fuselages and nacelles for the Boeing 737Max built by Spirit Aero Systems in Wichita, Kansas, are shipped by train across the country to Renton, in Washington State. The internationalization of aerospace manufacturing has become the norm rather than the exception in the 21st century (**Figure 5**). Aircraft wings made in the United Kingdom are shipped to Airbus in France and Germany and to Canada for the Bombardier CS jetliners, and executive and regional airliner fuselages are made in Mexico and also shipped to Canada. Fuselage and tail sections are made in Italy and Japan for Boeing final assembly, and many Western commercial aircraft are now part manufactured in China with final assembly there for the domestic market.

FIGURE 5 The Bombardier Regional Jet final assembly line incorporates major structures from the United Kingdom and Mexico as well as Canada.

FIGURE 6 CATIA digital tools have revolutionized design and manufacturing.

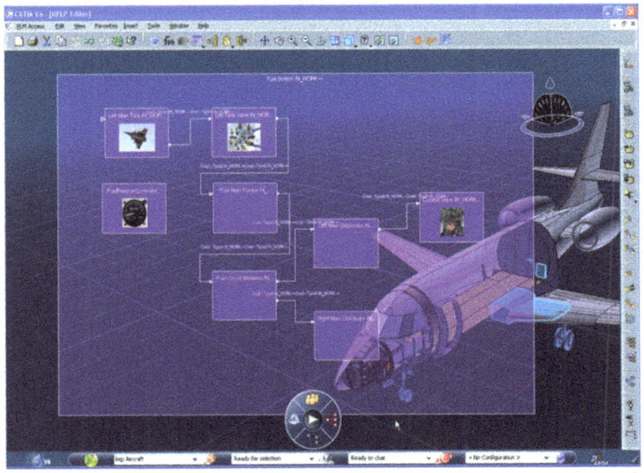

Digitally Enabled Manufacturing

Such diversity and common standards in the commercial aviation supply chain have been encouraged by the digitization of the whole design and manufacturing process through the availability of CAD/CAM programs (**Figure 6**). These digital tools provide distribution of test data and detailed specifications at every level of manufacturing activity and available to everyone who needs it. The industry-wide adoption of digital design tools was led by Dassault Systemes [9] whose family of CATIA products have revolutionized the way aerospace manufacturing is conducted. Virtual 3D simulation of assembly methods allows designers and engineers to see how best to route wiring and pipes, linking complex components around fixed airframe features, such as support struts and bulkheads, allowing visibility and access for repair or maintenance. Eliminating the need for making physical mock-ups to check that everything fits well, the virtual tools allow the entire aircraft design to be optimized and options evaluated before the production phase is started. The savings in time and effort are enormous. Highly automated computer-controlled tools build to extremely high levels of tolerance and consistency, which translates into higher quality and part traceability, an essential aspect of safety-critical aerospace manufacturing (Figure 7). The high level of non-degrading repeatability in automated production creates minimum wastage, which increases productivity and reduces material and factory operating costs. If holes are drilled or components machined or molded with even the smallest inconsistency, the results can be very expensive to rectify. Such essential environmental allowances are now factored into the manufacturing programs. Changes made by the design office on a new program using CAD/CAM technology can be communicated in real time to all those involved no matter where located. Advanced technology helps to design the aircraft and to optimize production line and assembly processes with all the systems integration and associated engineering requirements, including the use of robotic devices to make the whole end-to-end manufacturing structure as efficient and de-risked as possible (Figure 8). In place of large teams of aerospace workers swarming over semi-completed airframes on the assembly line, with sections being craned into position, using tooling towers to align them and work-station platforms for riveters and others to reach different access positions, on new programs, there is a move toward the major sections progressing along the assembly line on automated moving cradles. Multi-axis robotic arms carry out drilling and riveting as the fuselage sections come together in a steady stream. Automated arc welding further reduces assembly time. As the robotic processes become more flexible in functionality, much time and human labor can be saved by reducing the number of actions required to position and remove tools and parts.

FIGURE 7 Digitally controlled drilling.

FIGURE 8 Automated composite wing spar jigs.

New materials and manufacturing methodologies in aerospace are in a state of continuous development, and while the pace of innovation is increasing at an exponential rate, it can be a decade or more before the results of breakthrough technologies find application in actual production.

FIGURE 9 Complex printed parts can be designed and produced using advanced digital techniques.

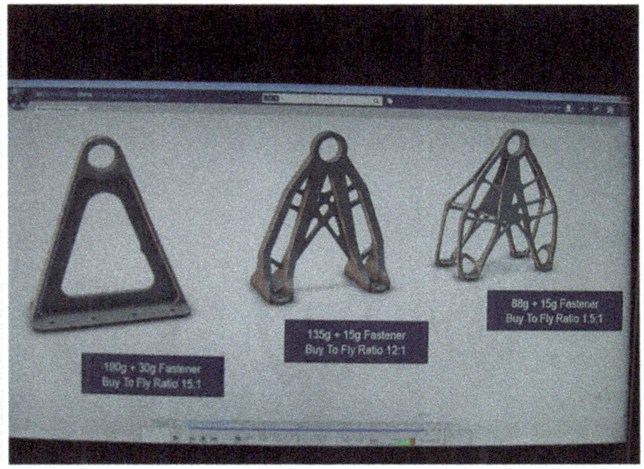

Examples of this can be seen across a wide spectrum, from the use of composite and hybrid materials in aerostructures and engine fan blades to additive manufacturing (AM) techniques with 3D printing of components. These materials and techniques within a safety-critical industry, such as aerospace, typically can take a comparatively long time to mature to the stage where they can be applied with confidence. Notwithstanding, this work has already been underway since the 1950s in the case of composite manufacturing, and for more than a decade with AM, utilizing advanced computer analysis programs to explore the parameters of what is possible. AM has tremendous potential for producing ultra-challenging shaped components and structures where casting, drilling, machining, or molding might not be possible, and a starting point in their use in the aerospace sector has been the manufacture of specialist components that are optimized for the requirement, but can be stronger, lighter, cheaper, and easier to make (Figure 9). However, like any other aerospace component, performance in extreme high or low temperatures, differing humidity levels, and under stress are just some of the factors that must be tested and proven acceptable before any new AM product can be certified for use. Some highly innovative component designs can only be made using this process. Many of the new manufacturing techniques are so potentially game-changing in manufacturing terms, they are kept well under wraps within the companies developing them, and in particular the major highly competitive engine manufacturers. AM does allow rapid prototyping, and because the entire process is computer controlled, exciting opportunities await as all-new programs can look forward to a much quicker design-to-test cycle of development. The route to everyday use of AM is thus a demanding one, but there can be no doubt that in the long term, the techniques will allow very unconventional components, structures, and perhaps complete integrated vehicle solutions, to be built. The latest integrated manufacturing processes involving AM and composites are moving toward the incorporation of more sensors and systems within "smart structures." The use of AM is growing from small components initially, but in composites, this is already seeing the incorporation of such features as anti-lightning protection and de-icing components being built into fuselage and wing sections. Less individual parts allow quicker assembly and reduced weight, and all such advances help reduce manufacturing costs.

The large-scale use of composite aerospace structures has rapidly expanded from such items as fuselage and wing panels, flaps, rudders, winglets, and cargo and landing gear doors, to complete primary structures—whole fuselage sections and wings, engine nacelles, and cabin and internal fittings. Complete airframes made from composites have emerged in the general aviation and executive jet markets, and an increasing proportion of larger commercial aircraft also incorporate many composite sections, the largest examples of major structural use being found on the Boeing 787 and Airbus A350 and A380 airliners. The composite manufacturing processes have been developed beyond the initial manual lay-down methods where impregnated sheets are placed on formers before entering an

FIGURE 10 Autoclaves are used in the manufacture of large composite structures.

autoclave to be cured in the molding stage. The latest digital design programs can specify and automatically lay exactly the optimum weave pattern and material depth required for a new composite part, taking into consideration how it will add strength to the shaped component, factoring in any special stress requirements. More automated methods are speeding up the composite manufacturing process, and multiple autoclaves are in use in the largest aerospace structure factories (**Figure 10**). Advanced metallic materials in the form of high-strength, low-weight aluminum alloys have made a comeback and are in widespread use as castings and machined parts. The special qualities of the latest aerospace-certified alloy materials can effectively compete in the form of single cast parts with more traditional fabricated alloy components, reducing weight and costs and also offering increased competition for composite suppliers.

The market has adapted to a changing aerospace business pattern, with more specialization by companies such as Spirit and GKN, who now not only produce specialist aircraft components but also deliver whole fuselages and wings for the world's leading aircraft manufacturers. This outsourcing of production by the major primes is growing, enabled by the use of industry-wide common CAD/CAM and certification standards, networked across continents and oceans, and has allowed a near-seamless global manufacturing model to emerge. The newest commercial programs, such as the A350 and B777-X, exploit the highest levels of automation in their construction and assembly. This automation is designed to greatly speed up the manufacturing cycle to deliver production standard aircraft with far fewer introductory issues than in the past. The spotlight has fallen in more recent times on the ability of suppliers to keep up with the demands placed on them by surges and reductions in production rates. Such items as in-flight communications and entertainment systems and cabin seats are increasingly sophisticated but are needed in very significant numbers to satisfy global airline customers who are responsive to competitive market expectations. Suppliers are under constant pressure from the major manufacturers on costs and delivery rates, yet have to invest in manufacturing capabilities on a scale that is flexible enough to effectively handle big demand swings. For smaller companies producing niche products, this is not always easy to manage and problems increasing production can impact disproportionally on aircraft delivery schedules. This also affects much larger suppliers as well, typically engine manufacturers, where a late-running new engine model can also seriously disrupt the contracted delivery to an airline, triggering compensation payments which might cancel profits on the deal.

The concept of modularity has a long tradition in aerospace manufacturing, but digitization methodologies have increased the extent to which its application can be more fully exploited. In the 1950s and 1960s, Boeing adopted much commonality on its model B707s, B727s, and B737s which shared many fuselage panels and similar cockpits. Airbus took this practice further with its wide-body family comprising the A300, A310, A330, and A340. Its single-aisle family introduced even higher levels of commonality on the A318, A319, A320, and A321, and this family shared a similar fly-by-wire equipped cockpit layout to the bigger A330 and A340.

The Powerplant Revolution

FIGURE 11 Sectioned model of a Pratt & Whitney Geared Turbofan engine.

The simplicity of the concept of the jet engine—suck (air), squeeze, burn, blow—has not changed, but today's very efficient, high thrust-to-weight turbofan engines have revolutionized commercial aviation, making possible ultra-reliable airline operations that even a few decades ago would have been unimaginable. Twin-engine commercial aircraft can fly continuously over oceans for 14 hours or more, and every day tens of thousands of aircraft fly safely across the globe, requiring very little maintenance. The construction of the biggest thrust jet engines is now so specialized there are only a handful of manufacturers in the world that are capable of making them and supporting them in service. All the major large global civil aircraft have Western-built engines, and even the latest Russian and Chinese manufacturers use them, or jointly produced engines. The main commercial suppliers are Pratt & Whitney, General Electric Engines, Rolls-Royce, and CFM International (GE and Safran). The coming of the wide-body designs in the 1970s drove the engine market toward a new generation of twin-spool turbofans that would combine very large low-pressure (LP) fans mated to the high-pressure (HP) compressors, creating the massive thrust required, but with the by-passed cold air from the LP fan mixing with the hot HP jet efflux to reduce noise levels very considerably. All manufacturers have continuously developed their own high-thrust engines incorporating new technologies as they have been matured. Fan blades have traditionally been made of metal, but composites are increasingly being used. Digital design and refining has produced extremely optimized fan blade shapes. Some of these "disruptive" technologies are guarded closely in such a competitive market and, in the recent past, have included growing components in crystal form and ceramic coatings, and even printed components. The transfer of new technology from R&D evaluation to full-scale production involves intense safety and capability testing which cannot be achieved quickly. Fan blades are now being made with hybrid composite and titanium. A relatively new development, not without its development problems, has been the "geared" turbofan, where a gearbox is inserted between the LP and HP stages to maximize efficiency at different speeds (Figure 11).

Managing the Supply Chain

The international manufacturing trend, for all the advantages it has brought, has nevertheless increased the dependency of primes on their supply chain. On a large commercial aircraft manufactured in the West, a majority of the specialists involved in bringing the program from design to manufacture will probably not be employed directly by the prime manufacturer, but will be located elsewhere in the supply chain. The arguments

FIGURE 12 Digital cockpit on the Boeing 787-9 features displays and avionics systems produced by specialist suppliers, but the company is planning to bring some of these capabilities in-house.

for and against self-sufficiency or outsourcing in aerospace manufacturing vary with different programs. Historically where primes have failed in the past and quality control and delays have arisen, this has been due largely to over-reliance on devolved supply management and not enough overall transparency across the whole program. Today, these incidents are rare, and digital tools allow all stakeholders in a program to remain appraised of its status throughout the development and manufacturing process in real time.

In 2017, the Boeing Company announced a significant change in policy direction regarding the supply of major systems and components, with the announcement that it would be creating a new in-house avionics organization to develop and build key aircraft systems (Figure 12). Ironically, for at least a decade both Airbus and Boeing have encouraged the Tier 1 and Tier 2 suppliers to consolidate to help them reduce the supply options for customers and to simplify the management of the supply chain, as well as taking advantage of the economies of scale. This request has been driven through with tough demands made on all suppliers to reduce costs in exchange for commitments to longer term production flows, and there is now evidence that at least some of this consolidation has been counterproductive for the big OEMs. By having fewer suppliers to deal with, the OEMs also have less choice in reaching contract agreements and the remaining suppliers have a strong negotiating stance. In order to comply with the tight cost targets demanded, the suppliers have had to reduce spare surge production capacity, and this has caused delivery delays for the OEMs as they have often had to wait as suppliers struggle to deliver increased production. The Boeing move to take much of this work in-house will involve a very big investment if they are going to introduce an alternative policy in time to adopt it for the next major program, which will be a new "Middle of the Market" wide-body to replace the Boeing 767, and as the Tier 1 and 2 suppliers have become very focused on their specialist areas of excellence. The international leaders that Boeing will be competing with include Honeywell, Thales, Rockwell Collins, and United Technologies Aerospace Systems (now Collins Aerospace).

Civil aerospace manufacturing has come a long way in just a century, and throughout this brief history, there have been many surges in innovation and progress, followed by periods of quiet evolution, before the cycle returns. Having experienced a thirty-year plateau where commercial airliners have not become any faster, or much bigger, just about everything else has changed beneath the surface, especially in the way the product is manufactured. The time is approaching, thanks to computer-designed and robotically constructed products, when the conventional winged tube airframe, with underslung engines, will probably become a museum item, as more exotic designs appear in manufacturers' brochures, offering even more efficiency, safety, and comfort, powered by hybrid engines, and where pilots are there just to monitor automated flight systems, and to provide reassurance for the fare-paying customers. And as manufacturing methodologies become ever more sophisticated, with perhaps whole aircraft being printed in 3D, or grown from elemental materials, with artificial intelligence taking over many decision-making tasks, the aerospace worker of the future will still be a vital element at the heart of the most challenging industry of them all.

References

1. Carter, C.F., RAE Report Aero 2150, p. 42, 1947.
2. Hartley, K., *Political Economy of Aerospace Industries*, (Cheltenham: Edward Elgar Publications, 2014), 12, ISBN: 9781782544951
3. Leary, W.M., *Encyclopedia of American Business History: The Airline Industry*, (New York: Facts On File, 1992), ISBN 0-8160-2675-0
4. Leary, W.M., *Encyclopedia of American Business History: The Airline Industry*, (New York: Facts On File, 1992.
5. Gardner, R., *Delivering the Dream*, (London: Royal Aeronautical Society, 2016, 29.
6. Tye, W., "Formation of the Air Registration Board," *The Aeronautical Journal, Royal Aeronautical Society*, November 1970, 877
7. Hartley, K., *Political Economy of Aerospace Industries*, (Cheltenham: Edward Elgar Publication, 2014), 12, ISBN 9781782544951.
8. Taylor, M., *The Aerospace Chronology*, (London: Jane's Publishing, 1989, ISBN 185-488-003-9.
9. Gardner, R., *Delivering the Dream*, (London: Royal Aeronautical Society, 2016), 105-107.

CHAPTER 5

Testing: What Works and What Doesn't

Richard Gardner
Farnborough Air Sciences Trust

Where can you get such an inexpensive computer, weighing only 200 lbs in full production the world over, and produced with entirely unskilled labour?

Captain Charles "Chuck" Yeager, the first pilot to exceed the speed of sound, speaking on the role of a test pilot

Designing and building an airplane capable of sustained flight is always an achievement, but manufacturing one in significant quantities that is reliable and safe for economical, everyday passenger carrying is an even bigger challenge. Such a machine must be subject to extended tests and evaluations, under different typical operating conditions, so that the element of risk to those on board, and on the ground, is estimable and minimized. But testing a new airplane is a more fundamental exercise than simply making sure that it is safe to fly. Testing will show whether the design is meeting, or failing, the expectations of the manufacturer and customers, a vital factor in the competitive civil market. It is also the proven route to correcting unexpected problems, and discovering new issues that must be addressed. In making improvements at an early stage of testing, time and money can be saved within the overall program and better solutions delivered. This chapter will examine the evolution of, need for, and extensive range of tests necessary for modern civil aircraft and their numerous systems, including the role of the human pilots.

Before any new model reaches the stage of development when it can be prepared for certification, every aspect of its construction and performance and every component and system within its airframe must be subjected to the closest examination. It also has to be tested to ensure that not only is the structure strong enough, but that it all works correctly and as expected, and with a sufficient margin of extra safety to standards demanded by the European Aviation Safety Agency and the Federal Aviation Administration in the United States. It needs to be capable of continuing to operate under pilot control in extreme environmental conditions, including severe turbulence, and despite performance degradation due to system failures or in-flight damage. In other words, it needs to be robust enough, with built-in redundancy, to fly safely in an emergency, which might be due to an unexpected airframe or engine fault, or even such unlikely conditions as flying through a volcanic dust cloud. Each component is subjected to very accurate process controls in manufacturing, and as 3D computer-controlled manufacturing tools have become an industry standard, it is now possible to guarantee manufacturing accuracy and consistency that allows large-volume production with confidence. With modern aircraft packed with complex electronic as well as hydraulic and mechanical systems and components, the interaction between all these parts must be thoroughly proven working within an integrated format to avoid electrical or mechanical interference.

The digital revolution brought a massively expanded role for software-controlled electronic systems, and this has introduced further technical challenges integrating multirole avionics systems and keeping them updated and user-friendly for aircrew and maintainers. With advanced computerized systems evolving far quicker than the airframes in which they feature, aircraft manufacturers are increasingly trying to "future proof" new aircraft designs with electronic systems with open architectures so that it is easier later on to upgrade models with new software and hardware packages. But while this trend now permits a more efficient way of improving performance, the importance of testing every aspect of such upgrades is no less essential, and can still be very challenging.

Static Tests

The Wright brothers in 1903 introduced powered flight as we know it, and they approached the problems associated with experimental work in a very scientific way. Although they were self-taught engineers, they nevertheless appreciated the need to study previous research and testing work carried out by other pioneer aviators. They realized at an early stage that building a small wind tunnel could help them experiment with and construct wing sections that would enhance lift while minimizing drag. Initially, the application of mathematical calculations relating to aerodynamics testing was geared more toward improving performance rather than safe handling of the aircraft. Much effort was expended in examining stresses in airframe structures with less on external load requirements. As aircraft got faster and higher flying, more attention was demanded in order to address fatigue issues, which had often remained largely hidden until accidents suggested a fault pattern. Pressurized civil airliners were introduced during the 1940s and 1950s, and this soon brought the need for a new look at how aircraft were being designed and built, as a consequence of repeated flights involving a rapid climb to altitudes well above the usual 10,000-15,000 ft, followed by many hours in the cruise at up to 30,000 ft, followed by a rapid descent for landing. With the extra stresses on the conventional metal airframes that now contained the pressurized cabins, new lessons had to be learned, and they did not come without a high price being paid in accidents. The most significant and high-profile series of fatal civil air accidents in the early 1950s occurred to the pioneering De Havilland Comet, the world's first jet airliner. The frequent cycle of operations at unprecedented high speed and at high altitudes represented a more than doubling in aircraft performance in terms of speed and productivity in a given timescale compared to the piston-engine airliners of the period. This placed additional loads on the airframe which in a relatively short period resulted in fatigue cracks developing at weak spots (roof hatches and window frames), and this caused explosive decompression and disintegration of the aircraft. Certification for the type was revoked and the growing fleet of new jets was grounded. Solving the mystery of what was causing these crashes became an urgent issue for the manufacturers, the airlines, and the British government. It saw a massive investigative effort by scientists and engineers which centered on building a large water tank at the Royal Aircraft Establishment at Farnborough, where a complete specially instrumented Comet aircraft fuselage was encased, with equivalent pressure exerted on the cabin as experienced during many flight cycles, until it failed (Figure 1). This provided clear evidence of how and where the structural failure was occurring and enabled the design to be modified with additional strengthening through thicker materials and revised airframe manufacturing features. The test results were distributed widely throughout the international aerospace community, and this provided a basis for the subsequent development of the first and subsequent generations of commercial jets.

FIGURE 1 A DH Comet jetliner being tested in the Royal Aircraft Establishment (RAE) pressure tank built at Farnborough to investigate structural fatigue issues in high-speed and high-altitude flight.

Fatigue testing in special rigs to explore structural limits and the robustness of design has been undertaken for many decades, and today destructive testing still features in the certification of new aircraft, where major structural components, such as wings, are subjected via hydraulic rams, to massive structural pressures, at an increasing rate, to the point where they break (Figure 2). Typically this happens only after they have been distorted to well over what might be expected in the most extreme loading conditions with turbulence. With the increasing use of composite primary structures in commercial aircraft, these physical tests have lost none of their importance, and results to date confirm, according to Airbus and Boeing, that composite wings are even stronger, and more flexible, than conventional all-metallic wings (Figure 3). Non-destructive testing and inspection is carried

FIGURE 2 An extreme test to destruction of a conventional all-metal wing.

FIGURE 3 An Airbus A350 composite wing structural test showing extreme deflection before failure.

FIGURE 4 A test section of a composite fuselage primary structure on a Boeing 787.

out to ensure material and structural standards are being maintained. The crack or flaw detection techniques include magnetic flaw detection, ultrasonic flaw detection, and the use of x-rays and laser detection equipment. Portable equipment is used to provide penetrative data, and a more recent development for visual inspections of large aircraft involves the use of small drone vehicles which can inspect the more difficult to reach areas of a wide-body airliner in a fraction of the time taken using tall scaffold and tower inspections.

The growing complexity of designs required more detailed testing and analysis of how key components performed individually as well as when combined in a system (**Figure 4**). This led to the building of test rigs that could replicate in realistic conditions linked working systems, including such features as fuel and hydraulic systems, electrical wiring circuits, power-operated actuators and flaps, and such additional features as anti-icing systems and landing gear retraction rigs. In recent times these system test rigs have become generally known as "iron birds" and are used to de-risk and test key systems and their interactions before they are fitted to the actual aircraft.

Complex Machinery

The engines are the most complex machinery in an aircraft, and with thousands of individual parts in each unit, testing is a constant feature throughout the cycle from design to manufacture and service. The latest commercial turbofans feature ever-advancing fan and turbine blade designs, with a mix of materials including nickel super alloys, titanium, and composites, requiring special manufacturing techniques (**Figure 5**). The individual turbine blades are hollow with complex passages for cooling purposes, and this presents particular challenges where conventional drilling may not be possible in manufacture: highly complex shapes can be "grown" using crystal-forming or additive techniques. Special ceramic coatings are also needed to protect engine blades from the effects of extreme temperatures above 1,400 degrees C. The latest high-thrust turbofan engines, such as the Rolls-Royce Trent XWB, can suck in 1.3 tons of air each second, and the high-pressure turbine blades rotate at 12,500 rpm, with blade tips reaching 1,200 mph. In regular service, these massive power plants, comprising over 20,000 parts, must function perfectly and continuously for up to 15 hours. The testing that is required before such a product can be certified for airline service takes years of intensive activity, measuring every parameter of performance from the robust qualities of an individual component to the ability of the complete engine to withstand damage in full-thrust conditions. Thermal shock tests are needed to see how components will perform when temperature changes occur within an engine. Ground-based test rigs are used extensively, fully instrumented to yield results that must confirm predicted performance and also indicate how successfully this can be sustained over different time cycles. Destructive testing can be spectacular and includes blade-off rig tests in which blades are deliberately detached at high-thrust settings to check on the debris containment properties of the engine case and nacelle, and their ability to withstand the destructive kinetic forces, and also to see the impact on the rest of the engine, which may or may not continue to part function, according to the certification requirements. Engines are also rig-tested for bird-strike evaluation, birds (proportional to the inlet size) being fired from a special gun immediately into the engine intake.

FIGURE 5 A test sample of an advanced composite and titanium fan blade.

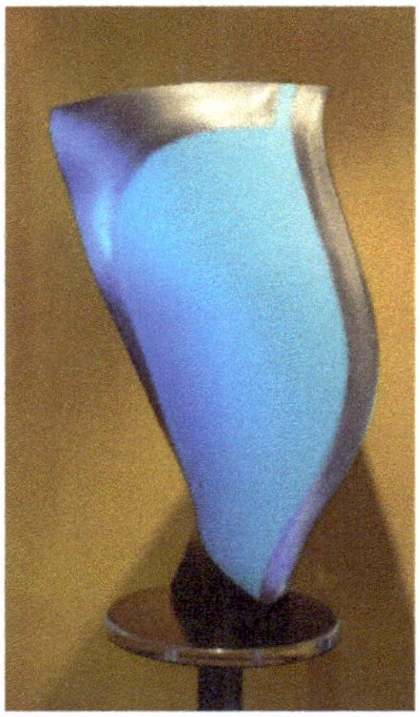

Environmental tests are also carried out using ground-based rigs to project simulated rain, hail, freezing rain, and ice into intakes, and for full engine tests in "cold soak" sub-zero conditions, and also extremely hot conditions. Full-size aircraft are also subjected to similar extreme environment tests in suitably equipped test hangars as well as at appropriately chosen geographical locations during the flight test program. These tests check how start-up procedures will work and also confirm that essential systems, including moving surfaces, aerials and sensors, doors, and access panels, also fully function. All new commercial aircraft also undergo evaluation in ElectroMagnetic Compatibility facilities where essential systems can be subjected on the ground to various electromagnetic pulse tests that ensure interference, such as from lightning strikes, or powerful ground emissions, will not disable the aircraft or its engines and systems, including communications, and that static discharges work as expected.

Another important ground-based test that all new commercial aircraft must successfully demonstrate before certification is for passenger and crew emergency evacuation. Using volunteers to represent full-capacity payloads, the required evacuation procedures are tested in realistic conditions and include the simulated lead-up period on board, with passengers receiving instructions to brace themselves, followed by timed and video-recorded observation of how effectively emergency doors and normal doors can be opened, escape chutes deployed and inflated, and passengers evacuated from all their seat rows, all within the specified timings, depending on the size and configuration of the aircraft. Changes may be required to internal arrangements, and even emergency door provisions may need to be changed, a very expensive modification, but only when the evacuation specification has been fully achieved will the aircraft be cleared for service. The internal design of new ultra-long-range civil aircraft, incorporating above or below deck crew rest areas, with some even proposing a limited number of above-cabin passenger sleeping areas, introduces additional safety issues that must addressed to protect users and to ensure that their access to cabin emergency doors is not impeded (**Figure 6**).

FIGURE 6 As well as being tested for passenger evacuation performance, every new aircraft is subject to evaluation at the design mock-up stage, as in this crew rest area.

New Procedures

Alongside digital test and evaluation tools have been more comprehensive and systematic fault analysis methodologies and procedures. Back in the 1940s the development of in-flight recording devices and analysis techniques enabled probabilistic methods to be devised to interpret the data results to give an estimation of the possible frequency of failure occurrences—once per X thousand flights [1]. Increasing the number of parameters that were being monitored during a flight led to the introduction of the first "black box" crash recorders in the next decade. In addition to helping to determine what might have caused an aircraft to crash, such in-flight data and voice inputs all add up to providing aircraft manufacturers with a very accurate indication of how a completed aircraft and its systems perform in the air. What was needed to allow for a safe margin in performance was a concept that would show the statistical variability of factors which affect performance, such as drag and available power, and the expected frequency of failure of key power units. Calculated incident probability became a major debating issue within the airworthiness community during the 1950s and pointed toward the adoption during the 1960s and beyond of more rig-testing (Figure 7). Despite the rapid growth in sophisticated computer-aided design and test tools, which can certainly highlight or eliminate many design problems at an early stage of the program, physical testing of partial mock-ups, or prototype aircraft, is still required to confirm predicted performance and to enable teams of specialists to analyze data captured by sensors monitoring all key systems and structures. It is a tribute to the capabilities of computer-based testing and analysis and advanced specification rig-testing that the first examples of today's new commercial aircraft can be built to a high level of pre-production standard by the time they are ready for their first flight, and in most instances only relatively minor modifications have to be made prior to full-scale production. The Airbus A350 is perhaps the outstanding example to date of an almost perfect development program devoid of any significant modifications or production line changes.

FIGURE 7 The full-size RAE thermal test rig was able to replicate Concorde service cycles to keep knowledge of the high temperature and stress effects on the structure always well ahead of the actual hours accumulated on the production in-service aircraft.

There is a potential downside however to the universal adoption of fly-by-wire (FBW) systems and glass-screen cockpit displays that allow aircraft to perform safely by the use of computers, with key flight data presented automatically as required or as selected by the pilot. Under some extreme environmental or maneuvering conditions, perhaps accompanied by multiple systems' failure, computer-induced pilot error can, and has, resulted in the loss of an aircraft and its passengers, as in the case of the Air France A330 which perished in mid-Atlantic after flying into a severe tropical storm on a direct flight from Brazil to France. Over-reliance on computer systems for safe flight in an emergency has become an issue that is now being addressed by intensive simulator training for pilots on the ground. However, it is clear that ever-increasing dependence on electronic systems aboard all types of commercial aircraft does require more situational awareness of what may be needed to overcome problems that affect aircraft operation. Safety-critical systems on all modern aircraft have built-in redundancy with back-up "fail-safe" systems and independent computer-controlled electronic architecture. Despite intensive testing in advance, nevertheless serious system failures do occasionally occur, or pilots misinterpret presented data. It is the attempt to eliminate these situations that is seeing much effort placed on even more comprehensive training and test regimes during development programs leading up to and including pre-certification flight tests. The interaction between the aircraft configuration design and flight testing and between pilot and the aircraft, the man-machine interface (MMI), has changed greatly over the last two decades, and some industry observers believe that during that period the flight test programs may have concentrated on achieving certification goals without fully recognizing more fundamental changing pilot training needs. The issue of pilot workload is always a major consideration working toward certification of a new aircraft, but many airlines were slow to appreciate how great a step it was for pilots to adjust from conventional aircraft to the new FBW generation, having been assured that the computer's programmed flight control laws would not allow the aircraft to exceed its normal flying parameters. Replacing the flight engineer in the cockpit with automatic systems controls, computer-provided information available to the pilot on multi-functioning glass screens, or as an automated action, was presented as a more user-friendly cockpit environment, reducing the complexity of routine operations. However, while the new FBW aircraft were built to be the safest ever put into production, flight testing cannot always allow for every possible unlikely combination of factors, including pilot error, with incorrect or self-cancelling data inputs, that could still put the aircraft at risk. Such human factors will always remain an outside possibility in anticipating error, but testing, at all levels, and personnel training, is aimed at reducing risk to the absolute minimum. New safety management systems seek to address this complexity. In due course increasing adoption of artificial intelligence systems will appear in new aircraft to replace, or anticipate, human decision-making in the cockpit. Already most routine piloting operations are automated, but this will inevitably expand and will assist future aircrews in dealing with flights in increasingly dense air traffic conditions and where artificial intelligence will ease even further aspects of multi-tasking that can produce aircrew stress. However, these developments will require a testing regime that demonstrates previously unattainable levels of systems infallibility to reassure airline passengers that it is fully trustworthy. Passenger resistance to flying in a pilotless commercial aircraft will almost certainly ensure that a pilot remains available in the cockpit, if only to monitor systems and provide assistance in an emergency.

Traditionally, wind tunnel models were used as a major source of data collection and analysis, the results of which could be scaled up mathematically to assist in calculating how a full-size aircraft would perform. Aerodynamics is all about the way that air moves and the reactions felt by surfaces to its motions. The wing shape and configuration can be tunnel tested to show stall characteristics, helping to determine minimum low speeds that can be flown, and how lift, stability, and controllability can be optimized, and drag reduced (Figure 8). Later, as more was understood about aerodynamic performance using wind tunnels, retractable lift devices, including leading edge slats and flaps, were tested to investigate how wings could provide even more lift by extending the surface area and optimizing flow at

FIGURE 8 Wind tunnel testing is as important today as in the days of the Wright brothers, and without it the Concorde supersonic transport could not have been brought forward to airline service in the 1970s.

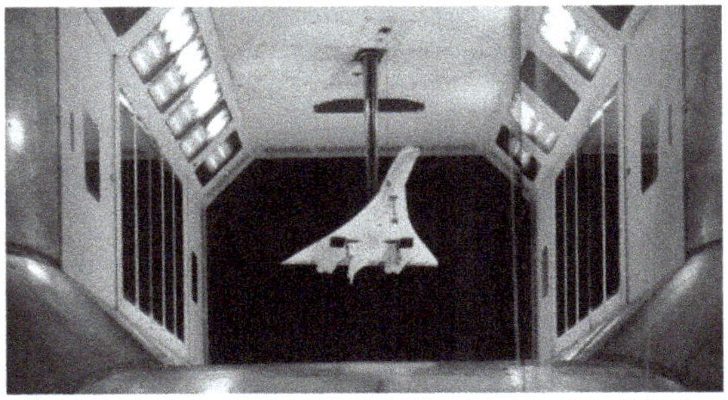

FIGURE 9 Wind tunnel test side view showing the effect of the fuselage design on transonic and supersonic shock waves.

slow speed without compromising the performance at higher speeds. Wind tunnel tests also allowed extensive study of how external aircraft features such as landing gear and external engine pods and pylons might interfere with airflow over the wings and tail. Another major use of wind tunnels was in the development of air intakes and jet nozzles and thrust reversers. Prior to the arrival of computer modelling techniques, including computational fluids dynamics (CFD), wind tunnel testing along traditional lines, involving embedded pressure sensors in the models, as well as smoke and fluid flows (used in conjunction with high-speed photography), streamed over a model mounted at different angles of attack and directional positioning, reached its zenith during the testing of supersonic and hypersonic aircraft designs (Figure 9). Conventional wind tunnel and prototype testing (extending down to key systems and equipment) has become extremely expensive and time consuming, and progress in computer modelling and simulation has revolutionized the risk-reduction aspects of developing new designs and has played the key role in keeping new programs affordable. With a fast growth in databases built up from extensive simulations and interpretation of results, combined with integrated design and manufacturing processes, and shared information in real time between development and manufacturing partners on an international scale, the overall task of testing a new aircraft design has not only reduced the development cycle but is also more transparent to all those involved in bringing it to production. Safety has lost none of its primacy in the process and indeed is fundamental to achieving certification, but physical testing of complete airframes and engines, as well as all systems such as landing gear and electronic and hydraulic systems, still produces unexpected surprises from time to time, with modifications required, but this is becoming rare. There are now an overwhelming number of opportunities during the simulation and computer modelling stages in development to eliminate emerging design, performance, or manufacturing problems and issues early on in the cycle, keeping delays to a minimum.

Today, wind tunnel testing is making a comeback in aerospace research and development programs, complementing CFD simulations and helping to de-risk advanced configurations, including disruptive blended wing/fuselage combinations. Both Airbus and Boeing have developed through advanced computer modelling unconventional blended-wing airframe designs that are inherently unstable but which maximize experience gained from military programs and extensive simulation to allow computerized flying controls to (potentially) deliver safe and reliable performance suitable for commercial operations. These designs have not yet emerged in any firm future product proposals and seem unlikely to be available for airline service until well into the 2030-2040 timeframe.

Why Flight Test?

According to the Advisory Group for Aerospace Research and Development (AGARD) originally formed as an agency within NATO in 1952, there are three reasons for carrying out live flight testing: to determine the actual characteristics of the aircraft (compared to the predicted characteristics), to provide development information, and to obtain research information. Aerospace is not an exact science and test results must be verified by measurement [2]. Even though computer modelling technology continues to advance to provide ever more accurate predicted data to confirm the mathematical model, it still cannot replace the verification made possible by wind tunnel and live test flights, using aircraft fitted with specially installed instrumentation. The stability of new designs needs verification at full-scale Reynolds numbers and Mach numbers, and aircraft sub-systems need to be tested in real-flight environments. Some flight testing can be carried out in special experimental platform aircraft, but in-development aircraft used for pre-certification flight testing, passenger seating is replaced by extra monitoring equipment and display consoles, observer and operator desks, and extra sensors and recording equipment (Figure 10). Each test flight results in huge amounts of data collected (including from ground tracking stations), and as well as being recorded for later analysis, much can also be shared with ground-based test personnel through real-time datalinks. Often the expected results are not the same as those predicted, but as a result of in-flight testing, design improvements can be brought forward and re-tested so the production aircraft can meet a performance level that is acceptable to the certification authorities and the customer. This is an expensive process and can show a better-than-expected performance, but can also lead to delays when problems arise, with customers possibly entitled to compensation as a result. However, these problems might not be severe enough to deny the certification award, and many commercial airliners have entered initial service in the full knowledge of the customer that improvement measures will be introduced over an agreed schedule. The Boeing 787 initially suffered in this way, the Airbus A320neo (re-engined) variants saw deliveries held up for many months, and the Bombardier C Series jetliner was also considerably delayed by engine supply deliveries and development problems. The future of the company itself, not just the program, could be at risk if the air-test phase and implementing its results takes considerably longer than planned. This might be caused by significant technical design, quality control, or performance issues, or it might be as a result of delays in the supply of key items of equipment, such as a new engine, but regardless of the cause, the outcome, if not resolved quickly, can result in considerable additional damage to a company's financial position, corporate reputation, and market position.

FIGURE 10 Performance monitoring aboard a Dassault Falcon 7X business jet on certification day.

The flight test program is a complex and interrelated process and has to be very carefully planned in order to achieve safety, costs, and schedules in balance [3] for an efficient and successful outcome. Each test stage will be examined by experts in the appropriate disciplines to consider performance and safety issues and their degree of severity and consequences. They will also look at the potential to reduce or combine the test actions needed to meet the requirements. Greater efficiency and test result productivity can be achieved when different tests are carried out on the same flight. This might involve technicians monitoring and recording the performance of on-board equipment and systems, while the pilots are testing aircraft handling performance at different altitudes

FIGURE 11 Tests in the air in fog and night conditions led to the certification of the Hawker Siddeley Trident as the first jetliner to be capable of automatic landing in all weathers in zero visual conditions.

and in different climatic conditions (Figure 11). Alternatives at each stage must be considered to obtain the data in the safest way. Because the performance of the aircraft is tested beyond what would be expected in normal service, there is always an element of risk attached, and so the test schedule and preparations must be accompanied by measures that guard against the consequences of an accident. This might include, for example, the provision of a stall-recovery parachute or a temporary fuselage tail-wheel to protect the rear airframe (typically on a stretched airframe) from damage during a "tail-drag" high-angle-of-attack take-off, and special flight crew emergency-egress arrangements.

As well as preparing for and taking a new aircraft through every stage of a test flight, to check out and explore specific performance and handling tasks, and under different visual and instrument conditions (day or night flying in hot, cold, and temperate climates), the test pilot must first go through a familiar routine representative of any typical flight, from flight planning, ground inspection, and pre-takeoff checks (fuel loads, payload weights, radio communications, engine, and aircraft systems). These activities will soon show up any shortcomings in the MMI ergonomics of the new cockpit controls design, though the use of realistic simulations and mock-ups should minimize such problems arising by the time the aircraft has reached the test-flight stage. Even though these human factor aspects of testing for a new aircraft can be comprehensively trialed on the ground, virtual environments alone cannot fully replicate the potential high stress that might be experienced when deliberate emergency recovery flying is tested in a real aircraft during a test flight. In-flight test requirements cover subjects including stall recovery performance, high- and low-speed handling, with and without the use of high-lift wing devices and air brakes, high angles of attack and banking, and engine switch-offs and restarts. There will also be tests landing on flooded runways to monitor water ingestion by the engines and air intakes on the airframe and landing gear bays. Extensive tests will also be undertaken to confirm predicted aircraft stopping capability at different weights and in different conditions from hot and high airfields and also runways in extreme cold climates, with and without the use of brakes, flaps, and reverse-thrust features.

FIGURE 12 Even before certification every new aircraft is shown off to potential customers at the major international air shows.

Asked what the role of the test pilot was, Bill Bedford, former Chief Test Pilot at Hawker Siddeley Aviation, said, "It is to observe and report on the safety of the aircraft to meet specification and to search for and eradicate unsafe features" [4]. Over the years thousands of test pilots worldwide have been involved in taking new aircraft up to, in many cases beyond, their designed operational limits, and many have perished in carrying out their essential work. But without their bravery, flying skills, technical experience, and fortitude, today's world of civil aerospace—so much taken for granted—simply would not have been possible (Figure 12).

References

1. Tyre, W., Airworthiness, *RAeS Aerospace Magazine*, March 12, 1975.
2. Ward, D.T., *Introduction to Flight Test Engineering*, Amsterdam: Elsevier, 1993.
3. Stinton, D., *Flying Qualities & Flight Testing of the Aeroplane*, Oxford: Blackwell Science, ISBN 063202121-7.
4. Bedford, A.W., "Role of the Test Pilot," *Journal of the Royal Aeronautical Society*, 68, 383, January 1964.

CHAPTER 6

The Journey of Certification: Safety Is Paramount

Hals Larsen
Propulsion Electronics, FAA - Retired Director

Ravi Rajamani
drD2 Consulting

> *Flying is so many parts skill, so many parts planning, so many parts maintenance, and so many parts luck. The trick is to reduce the luck by increasing the others.*
>
> David L. Baker

The certification of new items and systems is an important part of the process of introducing them into service. This is done via the process of obtaining a type certificate (TC) that the governing regulatory agency issues to an aircraft original equipment manufacturer (OEM) when the latter is able to demonstrate that the design of the aircraft is safe. To be able to produce the aircraft in quantity, the OEM also obtains a production certificate that verifies the manufacturing processes can ensure that the design will be faithfully rendered into a product.

However, having a TC is not enough to operate the aircraft. Appropriate regulatory authorities in the region the aircraft is based, who might be different from those that awarded the type certificate, have to be convinced that the operator has the ability to maintain the aircraft to ensure continued airworthiness throughout its lifetime. The operator may have its own maintenance shops, or it may outsource this function to a third party, but in either case, a rigorous maintenance process needs to be established so that design safety continues to be maintained.

In this chapter, we provide a brief overview of the certification process. The information is decidedly slanted toward the practices prevalent in the United States since one of the authors had many years of experience with the Federal Aviation Administration (FAA).

Types of Certificates

Certification of systems is governed in the United States by Federal Aviation Regulations (FAR) and falls under Code of Federal Regulation (CFR) Title 14. Different types of systems and components come under different parts of these regulations [1]. For example, small planes—carrying less than 20 passengers and having a takeoff weight of less than or equal to 19,000 lb—are governed by Part 23. Most general aviation aircraft fall in this category. Bigger transport aircraft, generally the ones most of us are familiar with as airline passengers, fall under Part 25. Because of their criticality in the function of flight, certain components within the aircraft need to get their own independent certificates—the engine under Part 33 and the propeller under Part 35. Rotorcraft fall under Parts 27 (small) and 29 (large), and, with the increasing popularity of unmanned aerial vehicles, certain bigger systems need to be certified as well under Part 107. This last category continues to evolve as the forms and functions of this class of aircraft evolve [2].

The laws of different countries differ in subtle ways as to how they honor each other's certificates. Because the FAA is probably the oldest and the most well-established of the various agencies, aircraft certified in the United States generally are recognized as such in other countries via reciprocal arrangements, and vice versa, e.g., the agreement between the United States and Canada [3]. In the past, Europe had independent civil aviation authorities, which cooperated under the Joint Aviation Authorities (JAA) guiding principles, but in 2002, the European Aviation Safety Agency (EASA) was established by the European Union to act as the single overarching agency to cover all of its member countries [4]. In addition, the International Civil Aviation Organization (ICAO) was created in 1947 [5] as an agency of the United Nations (UN) to establish uniform standards across all countries in various areas of civil aviation. ICAO does not interfere with individual authorities on certification and continued airworthiness issues.

In each of the categories above, when the manufacturer has demonstrated that its design can safely perform all flight functions according to the certification standards in 14 CFR, the authority will grant it a type certificate (TC) for that specific type of aircraft. By demonstrating that it has the ability to manufacture the product, the OEM can get a production certificate (PC). After the aircraft has a TC, if the original manufacturer, or any other entity with demonstrated skills, needs to modify the aircraft or any of its components, it has to apply for and obtain a supplemental type certificate (STC). The STC, which is related to the applicable TC, needs to outline how the modification affects the original design [6]. The FAA also publishes Technical Standard Orders (TSO) that outline the minimum performance standards for aircraft components and material. Any part designed and manufactured according to the TSO is guaranteed to be safe when used on an aircraft according to acceptable operational practices. However, before a company holding a TSO authorization (TSOA) can install a part on a specific aircraft, a TC or an STC is needed [7].

The TC and the PC ensure that the aircraft is designed according to certification standards and that the OEM is capable of producing the aircraft according to the design. These are typically granted to the OEM. To ensure that the aircraft is capable of actually operating under normal and abnormal conditions, airworthiness certification needs to be obtained by the operator.

To help guide the design teams, the regulatory authorities often point them to standards or guidance documents written by standard-setting organizations, such as the Society of Automotive Engineers (SAE International), via publications known as advisory circulars (AC). For example, the Aerospace Recommended Practices (ARP) ARP4754A [8] and ARP4761 [9] are the documents most organizations use to structure the certification process for any airborne equipment. ARP4754A lays out the overall systems engineering process for developing the system. This includes how requirements should be written, how System Safety Assessments must be carried out, what testing needs to be done to demonstrate the design functions, how systems should be validated and verified to ensure safety, etc., while ARP4761 lays out in detail the processes for doing the safety analysis. The safety analysis includes failure modes and effect analysis (FMEA), functional hazard assessment (FHA), fault tree analysis (FTA), etc. Another organization that is very active in this field is the Radio Technical Commission for Aeronautics (RTCA). It has developed a number of documents that are invaluable in the certification process such as DO178C [10] and DO254 [11] which deal with onboard software and airborne electronics, respectively, and DO160G [12], which details the environmental conditions under which all airborne equipment, including electrical equipment, need to show functional integrity. Figure 1, taken from ARP4754A, shows the relationship between the different documents. DO160G is not shown here since it deals with design at a more detailed level.

While it is a regulatory authority such as the FAA that is ultimately responsible for issuing certificates, the truth is that there just aren't enough people within these agencies to do all the work that needs to be done to obtain and maintain these certificates. The FAA, for example, has created a process by which organizations and individuals outside the government can obtain authorization to act on behalf of the regulatory body to help with this work. There are a number of such delegations currently authorized including Designated Airworthiness Representatives, Designated Engineering Representatives, Organization Designation Authorizations, etc. To learn more about these, the reader is directed to the FAA website [6].

Process of Obtaining the TC

The trick is to start the process as early as possible. That gives the responsible regulatory agency time to become acquainted and familiar with the items and systems before

FIGURE 1 Guideline documents covering development and in-service/operational phases (ARP4764A).

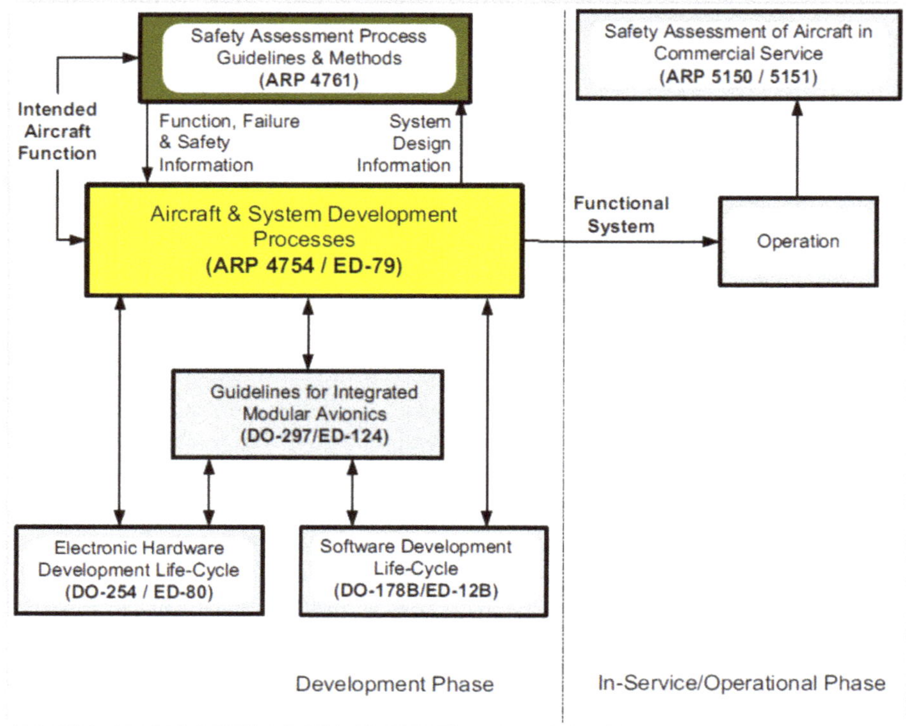

introducing them into field service. The key to the certification process is being clear on what the system or item functions are. A major goal of certification is to determine the myriad ways in which a system may fail, develop design or operations mitigation for these failures, and convince the authorities that everything has been done to make the risk of failure negligible. When certifying a system, important items to consider are the failure modes and their effects on the system. How do those system failure modes affect the aircraft, how are they apparent to the flight crew, and what should be taught to the flight crew concerning failure-mode recognition and their expected (thought and learned) responses? This exercise will involve bringing the flight crews into the process as necessary to explain and understand system operation and expected failure mode responses. If the failure modes and expected flight crew responses are not acceptable to the flight crew, the prudent course of action is to redesign the item or system, as necessary.

Several interested organizations have jointly published a guide to the certification process that can be consulted as an introduction to the topic [13]. An illustrative figure taken from this reference (Figure 2) shows how the certification process aligns with the typical product development cycle.

Ultimately, the most important aspect of certification is ensuring aircraft safety. If only certifying an item in a system, the failure modes of the item and its effects on the system as a whole need to be analyzed and mitigated. The safety assessment process will include discussions and documentation of those failures and failure modes and the

FIGURE 2 Typical project phases aligned with the certification process.

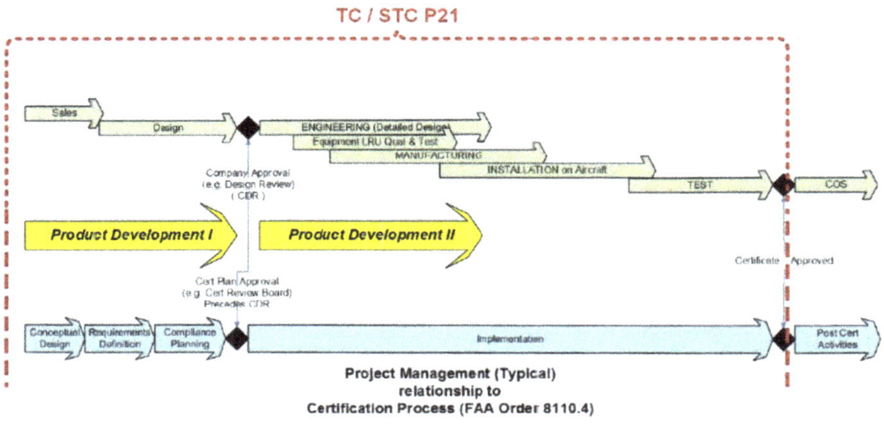

results on the system. The key to successfully obtaining a TC is to make sure that everything is documented with proper traceability, so that nothing, as far as possible, is left to the imagination and implication.

In addition to initial certification, the regulators have additional responsibility to ensure that any design flaws or other failures that are uncovered during operation are dealt with in the appropriate manner. After investigating such incidences, they typically issue airworthiness directives (AD) to correct problems.

Once the TC and PC are obtained, the aircraft is now in the hands of the operator. An airworthiness certificate is issued to the operator to allow the aircraft to be flown. To ensure continued airworthiness, the aircraft must be maintained properly according to OEM recommendations and standard industry practices. An industry consortium called the Maintenance Program Industry Group (MPIG) publishes sets of codified maintenance guidelines, now in its third version (MSG-3) which the FAA [14] has endorsed as the recommended approach to maintaining continued airworthiness of transport aircraft. The operator may perform these maintenance tasks internally or outsource them to certified third parties.

The regulatory authority maintains a Maintenance Review Board (MRB) for each type of aircraft and power plant to provide advice on how the system will be maintained. An informative flowchart, taken from AC 121-22C [14], lays out the process of establishing and conducting the MRB process (Figure 3). For even more details of how organizations in Europe and around the world maintain safety standards, the reader may consult the very informative website called SKYbrary [15].

Testing to Support Certification

Typically, the most convenient way to demonstrate safety during flight is to physically test the aircraft or item and prove to the authorities that it is airworthy. Because of the cost and time involved in these tests, only a limited number of tests can be done.

FIGURE 3 Maintenance Review Board process flowchart.

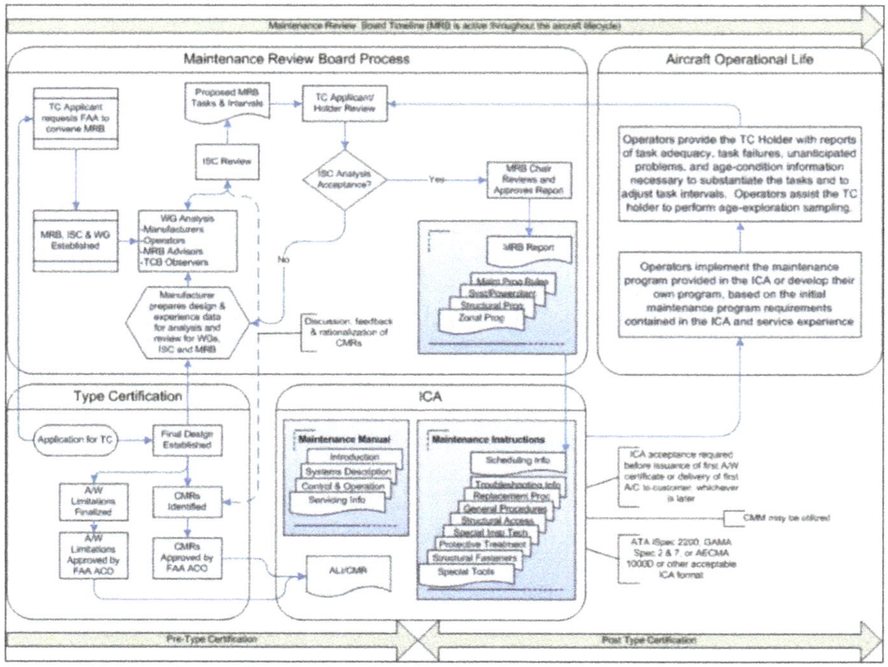

Some analytical techniques are used today to demonstrate design limits, but there is a growing realization in the industry that these analytical model-based techniques can be used even more extensively to demonstrate safe operation over wider flight conditions. We are going to see an increased use of analytical techniques such as simulations in the future during certification exercises [8]. Similarly, formal methods will become even more prevalent to demonstrate the safety of software systems in the future [16].

There are a series of tests performed during the certification period. While the final flight test is what demonstrates the entire system, many intermediate tests can be used to show compliance with different requirements in 14 CFR Part 25. For example, many electronic systems on the aircraft will have to be shown to comply with immunity to external radiation, and lightning strikes in addition to mechanical disturbances, which is done at the component level. Functional validation is a very important part of the system development process. This shows that the system that is being built is the right kind of system. In the past, it was very common to build prototypes to validate functional requirements, but the current practice is to rely more and more on model-based validation. Once validated, especially for safety considerations, the system can be built with more confidence. Some functions cannot be validated analytically, and for these other means, including prototypes, must be built and tested. System verification, which ensures that systems are built right according to the requirements, is a relatively more straightforward exercise.

The main guide that FAA publishes to guide applicants with the flight test is AC 25-7 version C [17]. Before the final certification test, the aircraft will go through many flights to test out various systems and subsystems, and it will be flown to its limits to cover the entire design envelope. Some typical tests for this include engine-out takeoff tests, brake effectiveness on a wet runway, and stall margin tests. The intent here is not to go into details of how tests are conducted to show compliance, but suffice it to say testing is an integral, critical, and expensive part of aircraft and engine certification.

Engine as an Example

In this section we include the example of an aircraft engine, to bring out some key features about its certification. The engine is one of the most critical components of the aircraft; its continued operation is crucial to the safety of the system and the mission. In addition to the physical turbine, the engine consists of a number of ancillary components of which the most important among them is the engine controller. In the case of engine control systems, the controller can be located in an aircraft electronics bay (e-bay) in the aircraft or on the engine. On the engine, the fan case is the most common location. Although it is considered a rather harsh environment with its wide temperature variations and vibration signatures, it has the convenience of being close to the engine, which makes the wire runs from the controller to the engine sensors reasonably short. In addition, with the controller located on the engine, pneumatic interfaces between the controller and the engine are more easily accommodated. There are other important engine control signals that come directly from the aircraft including the throttle position sensor, fuel shutoff command, aircraft ambient condition, airspeed, aircraft total temperature, altitude, etc. There are aircraft where the controllers are located on the aircraft, such as in the aft e-bay. In such configurations, attention should be focused on "common cause" threats such as e-bay fire or overheat events. Automatic detection and suppression of such threats is normally needed, as the loss to all engine controllers is considered catastrophic to most aircraft. In two-engine aircraft, a high-energy fan blade or engine disk failure can cause liberated engine parts to pass through the aircraft and strike the other engine. With four engines on a swept wing design, the chance of an engine disintegration taking out the engine controller of another engine on the same wing is very low. But, as was demonstrated on an Airbus A380 incident in 2010, high-energy debris from one engine caused control lines in the wing to be ruptured so that the other engine could not be shut down even though it was not physically damaged [18].

A high-vibration environment also has its challenges. With solid-state electronic devices, the vibration environment may not be that harsh for the electronic components themselves, but the boards on which the devices are located can be of concern. The boards can be more affected by vibration than the actual devices themselves, so the boards may need enhanced support.

Increasingly, aircraft manufactures prefer to hand over the responsibility of the entire engine module to the engine OEM. In the past, airframers took on the responsibility for integrating many parts of the engine at their assembly lines, but today, they prefer to buy an engine with all of its parts located on the engine and simply install it on the aircraft. This means that the engine OEM is responsible for everything that hangs off the pylon under the wing. They are responsible for conducting the certification activities associated with all these components.

The common threats to engine controllers are generally heat and vibration. External environmental factors such as volcanic ash can cause travel disruptions because of their effect on engine operation [19]. During the certification process, all these factors need to be taken into account to ensure that the necessary mitigation functions are investigated to ensure that the effects of the threats are minimized to an acceptable level.

The certification of items and systems is always a significant task. In the case of engine controllers, they actually go through two certification processes. The first is at the engine level, which is quite extensive. The controller must live through all expected engine disturbances. As mentioned above, one of the largest disturbances is vibration. Some of the more significant vibration levels can be those caused by the ingestion of large birds, ice shedding, or the loss of a fan blade. Thus, all testing related to these events are conducted with the controller in its normal location. These are some of the most severe tests involved in engine certification, and the controller needs to survive them without malfunctioning during the event.

Since the advent of full authority digital engine controllers (FADEC), the item of greatest cost and concern is the controller's software. Controller software essentially runs the entire engine from start-up to shutdown in all experienced environmental conditions, including restarting the engine on the ground and in the air, and under normal and adverse conditions.

Since the software is common to all engines, a fault in the code will affect all engines in that type. Because of this and because a failure in the engine software that can cause a catastrophic disaster, engine software is considered level A, the highest level of software integrity. Software design assurance levels are categorized into five levels—A through E—based on the effect could they have on the safe operation of the system they are controlling in case of failure. The detailed guidance on how to develop software at each level is given in DO178C [10]. The software development process is quite complicated, and it is necessarily so because of the criticality of the functions it supports. To ensure that software functions are as foolproof as possible, level A and B systems have to be designed with independent teams, so that there is a higher probability that mistakes are caught before they are implemented. For example, one set of engineers is responsible for developing requirements for level A software, while an entirely different set of engineers is employed to validate and verify it. Because of the nature of software, regulatory authorities tend to be more careful in its review. They conduct several reviews called stages of involvement reviews during the development of critical software, with the level of scrutiny appropriate to the level of criticality.

Control software is particularly critical because it covers engine start-up, running the engine and shutdown at all engine operating conditions. In addition, it has to operate the engine during all non-normal operating conditions as well, such as lower airspeeds than the engine is designed to operate. In general, the controller should never cause an engine to stop operating unless specific safety conditions necessitate such action.

These days, one of the most significant items that the software covers, and is expected to cover, is unexpected engine problems, such as engine damage. Even if the engine is damaged, the software is expected to run the engine, if engine operation is possible. If the engine is damaged, but can run, the controller should be designed to keep the engine running, albeit at a reduced capability depending on the health of the engine.

Engine controls now contain functionality to implement engine health management, which has not traditionally been part of the certification process. When an engine starts to deteriorate, the changes in its performance can be detected by the controller, and it can be operated at degraded conditions, obeying acceptable physical limits. Accordingly, a part of controller software programming is devoted to engine diagnostics. Engine controllers can also be programed to advise on maintenance actions based on analysis of sensed parameters. The alternative is to get a pilot's report of the engine problem and the scheduling of an unexpected/unscheduled engine maintenance activity, which is a costly item in today's operations. Thus, controllers are preprogrammed to forecast needed engine maintenance to the greatest extent possible and practical. These reports do not change scheduled engine maintenance. They help with needed unexpected maintenance. So, a great deal of engine software is devoted to engine health and diagnostics. This is obviously not just true of engines. Increasingly, many aircraft subsystems depend on software and the advanced capabilities it engenders. The next big revolution in this area will be the use of engine health management systems to replace scheduled engine maintenance actions. This is called obtaining engine maintenance credits and is the area of active research these days. Both the certifying authorities and those responsible for continued airworthiness will have to get involved in granting these credits. The process of certifying these functions will not be fundamentally different from the process for certifying normal functions in an engine but will require a lot more work on the part of the applicant because of its novelty.

The two major operating costs of an aircraft—other than personnel cost—are those associated with fuel and maintenance. Many large operators write their own maintenance manuals and instructions after getting familiar and accumulating experience on operating an aircraft. Although the OEM is obliged to provide maintenance manuals as a starting point, these are quickly revised by the operators. Even when maintenance is assigned to a third party, this activity is closely watched and scrutinized, so that maintenance costs can be minimized for the sake of the bottom line. The use of automating the diagnostic process and the increased use of analytics in this area can be very welcome, if the cost of certifying them is not overwhelming.

In summary, the maintenance of an engine and its controller is one of the higher cost items in operating an aircraft. Hence, the software in engine controllers is very much associated with containing the necessary and advisable maintenance instructions and related information. This becomes an important part of controller certification and approval. Since the FAA Flight Standards group is responsible for aircraft maintenance, approval of controller software not only involves the certification group but also the Flight Standards personnel. This makes it an interesting, albeit a sometimes frustrating, task. But with persistence and with the backing of analytics and validation data, even new engine health management functions can be approved for use on civil aircraft, thereby lowering the overall cost of operating the fleet and, at a future date, even enhancing flight safety.

References

1. Wikipedia Contributors, "Federal Aviation Regulations," Retrieved August 2017, https://en.wikipedia.org/wiki/Federal_Aviation_Regulations.
2. Federal Aviation Administration, "Fact Sheet – Small Unmanned Aircraft Regulations (Part 107)," June 2016, Retrieved August 2017, https://www.faa.gov/news/fact_sheets/news_story.cfm?newsId=20516.
3. FAA/TCCA, "Implementation Procedures for Airworthiness," Rev 2, November 2016.
4. SKYbrary Contributors, "European Aviation Safety Agency (EASA)," Retrieved August 2017, http://www.skybrary.aero/index.php/European_Aviation_Safety_Agency_(EASA).
5. International Civil Aviation Organization, "History of the ICAO," Retrieved August 2017, https://www.icao.int/secretariat/TechnicalCooperation/Pages/history.aspx.
6. Federal Aviation Administration, "Designees and Delegations," Retrieved January 2018, https://www.faa.gov/other_visit/aviation_industry/designees_delegations/.
7. Federal Aviation Administration, "Technical Standard Orders (TSO)," Retrieved August 2017, https://www.faa.gov/aircraft/air_cert/design_approvals/tso/.
8. SAE International, "9 Guidelines for Development of Civil Aircraft and Systems," ARP4754A (EUROCAE ED-79), December 2010.
9. SAE International, "Guidelines and Methods for Conducting the Safety Assessment Process on Civil Airborne Systems and Equipment," ARP4761, December 1996.
10. Radio Technical Commission for Aeronautics, "Software Considerations in Airborne Systems and Equipment Certification," RTCA DO-178C (EUROCAE ED-12C), December 2011.
11. Radio Technical Commission for Aeronautics, "Design Assurance Guidance for Airborne Electronic Hardware," RTCA DO-254 (EUROCAE ED-80), April 2000.
12. Radio Technical Commission for Aeronautics, "Environmental Conditions and Test Procedures for Airborne Equipment," RTCA DO-160G (EUROCAE ED-14), December 2014.
13. AIA, AEA, GAMA, and the FAA, *The FAA and Industry Guide to Product Certification*, 3rd Edition (Washington: The Aerospace Industries, May 2017).
14. Federal Aviation Administration, "Maintenance Review Boards, Maintenance Type Boards, and OEM/TCH Recommended Maintenance Procedures," Advisory Circular, AC 121-22C, August 2012.
15. European Organisation for the Safety of Air Navigation, "SKYbrary, The Single Point Reference for Aviation Safety Knowledge," Retrieved August 2017, https://www.skybrary.aero/.
16. Radio Technical Commission for Aeronautics, "Formal Methods Supplement to DO-178C and DO-278A," DO-333 (EUROCAE ED-216), December 2011.

17. Federal Aviation Administration, "AC 25-7C - Flight Test Guide For Certification Of Transport Category Airplanes," October 2012.

18. Wikipedia Contributors, "Qantas Flight 32," Retrieved August 2017, https://en.wikipedia.org/wiki/Qantas_Flight_32.

19. Eurocontrol, "Ash-Cloud of April and May 2010: Impact on Air Traffic," June 2010, Retrieved August 2017, https://www.eurocontrol.int/sites/default/files/article/attachments/201004-ash-impact-on-traffic.pdf.

CHAPTER 7

Operators: What Makes Airlines (Well) Airlines

David Piotrowski
Delta Air Lines

Running an airline is like having a baby: fun to conceive, but hell to deliver.

C.E. Woolman, Founder of Delta Air Lines

Introduction

The civil airline business is not for the faint-hearted. Massive logistical complexities with uncontrollable variables, thin financial margins, high capital costs, burdensome regulations, and thorny labor relations are all characteristics of the business. The best laid plans can be obliterated by weather, labor strife, air traffic issues, geopolitical confrontations, economic "shocks," oil pricing and supply, exuberant regulation, and aging infrastructure issues like power outages or water pipe bursts.

From baggage handlers, fuelers, and catering to maintenance, gate agents, flight attendants, and pilots, the complex operation hums like a well-oiled machine. Except when it doesn't. And it doesn't a lot. Time pressures, passenger volumes, and different personalities all mix with unexpected situations like weather, volcanoes, labor protests, geopolitical spats, mechanical breakdowns, and terrorism, to create delays, diversions, and cancellations.

High load factors mean that there is less margin for recovery, and the operation can be fouled up for days. The consequence is that the airline not only has passengers, baggage, and cargo in places where they don't want to be but also planes, pilots, and flight attendants in different locations than planned. So, restarting an operation after a disruptive event is not trivial.

If service is delivered poorly, or flight delays and cancellations result, then customers may be unsatisfied and elect not to fly with the airline in the future. This is important with the advent of social media, where any airline missteps are documented and posted online instantly. Airlines have established quite a poor reputation and are one of the most hated industries in existence [1]. An airline operator could even go out of business due operational failings and the resulting poor reputation.

Indeed, history is full of airlines which either failed, or were merged with other carriers, or had assets acquired by stronger carriers: Pan Am, TWA, Laker, British Caledonian, British European, Union de Transports Aériens (UTA) (France), Air Outre-Mer (AOM) (France), Air Berlin, Swissair, Societé Anonyme Belge d'Exploitation de la Navigation Aérienne (SABENA), Magyar Légiközlekedési Vállalat (MALÉV) (Hungary), Eastern (United States), Ansett Australia, Northwest, Continental, and U.S. Airways. However, there are a variety of airlines worldwide which are quite successful. Why do some thrive, while others do not manage to navigate these challenges? What does it take to run an airline?

Passenger Journey

Let's start by taking a look at the journey of a typical passenger. A passenger can purchase a ticket for a flight between 330 days and an hour prior to the flight. They may have done so through an airline website, other travel websites via global distribution systems, at a city ticket office, or a third-party service such as travel agents. The ticket is offered in a variety of forms, dependent on seat size and class of service, date of travel, time of day, destination, and may or may not come with a baggage allowance, entertainment, seat assignment, no change fees, and lounge access.

Next, the passenger heads to the airport on the day of their flight where they check-in, deliver their baggage, and undergo security screening. New technologies like online check-in and baggage delivery services mean the passenger may not even encounter an airline employee. Then the passenger boards their flight and is flown to their destination. Once onboard, the passenger could be offered additional products like Wi-Fi, entertainment options (movies, shows, satellite TV, songs), food, beverage, headsets, blankets, and pillows, which may or may not come with additional fees. Assuming smooth operation, the passenger deplanes and waits to collect the baggage, and also waits for immigration and customs queues (if necessary), before going on their way.

The passenger does this "routine," some so frequently that it becomes "common-place," yet may not know all that it takes behind the scenes. And it truly takes a carefully conducted team to execute this operation, even more so when things go wrong.

Business Strategy

Airlines must formulate and adhere to a strategy, which then determines which passengers to target, routes to fly, schedule to fly, aircraft to purchase, and what products to offer. Carriers spend great amounts of time analyzing traffic flows, economic indicators, and potential competition on their networks. This often formulates the overall strategy of whether to be a low-cost carrier, cargo carrier, niche carrier, leisure carrier, premium carrier, or a full-service network airline. Each of these types of carriers must comply with their appropriate regulations (i.e., in the United States, the Code of Federal Regulations, Title 14, Parts 121, 91, and 135) [2]. Some airlines around the world are utilized as a "public service" to ensure all parts of the region or country are served.

Each flight departing with empty seats has "perished inventory," as that seat at that time cannot be sold again. This process repeats every few hours as an aircraft makes several flights per day. Contrast this to a manufacturer of canned goods, whose product sits on a shelf in a store until someone purchases it. This affects the pricing via supply and demand, made more complex via different fare categories and different products. Thus, airlines only put their high-cost assets (aircraft and labor) where and when they can make money. Note that this is not necessarily the same as filling a plane, as plenty of routes become unprofitable if fares are too low.

Route Networks

Arguably, the most important item to the success of an airline is "where" to fly, along with "when" and "how often." Some carriers specialize in point-to-point routings, like easyJet, while others focus on leisure traffic, like Thomas Cook Airlines. You'll often see plenty of carriers clamor to offer service between London and Paris but fewer carriers between smaller markets like Glasgow and Naples.

The biggest carriers tend to develop a "hub-and-spoke" system in order to economically serve a larger variety of markets. Pioneered by the U.S. carriers, well-placed hubs allow them to collect traffic from a variety of points and send them onto other points [3, 4]. For example, the market between Harrisburg, Pennsylvania, and Omaha, Nebraska, is so small that no carriers fly the route nonstop. But passengers can get there via Detroit on Delta Air Lines, or through Chicago on United Airlines, or through Dallas on American Airlines, plus other hub options. These airlines are successful because they carry not only the Harrisburg passengers to Omaha but also passengers from Boston, Buffalo, Cincinnati, Raleigh, and Grand Rapids to Omaha. Multiply this route times hundreds of cities and thousands of flights and the economic efficiencies are powerful, providing benefits to airlines, passengers, and communities. **Figure 1** shows part of the new Terminal 5 for British Airways' hub at London Heathrow.

Network Planning

For the airline, choosing "how often" and "when" to fly to a city is actually not as easy as it seems. International flights must have rights or permissions to fly to and/or overfly a country. Russia makes significant money from air carriers flying through their airspace

FIGURE 1 British Airways has a hub operation at London's Heathrow Airport.

to Asia and Europe. Still other areas of the world are off-limits due to military conflicts and geopolitical standoffs, which also impact cost.

Some countries have negotiated reciprocal "Open Skies" agreements, allowing unlimited flights between carriers of two countries. This truly allows the free-market system to work, providing great economic benefit to communities, passengers, local businesses, and airlines. The United States and Europe are perhaps the largest example of the "Open Skies" model, and traffic has nearly doubled since the early 1990s [5]. Additionally, this increased the number of cities/hubs to receive flights. Instead of most transatlantic flights starting in New York, now there are transatlantic flights from Atlanta, Charlotte, Detroit, Denver, Austin, and Raleigh, among others. The same is true of Europe—instead of just London, Paris, or Frankfurt, cities such as Dusseldorf, Porto, Vienna, Prague, Brussels, and Milan all boast transatlantic flights.

However, many countries restrict the number of flights, number of seats, or even the fares in the markets. Emirates is restricted to only 3 daily flights to Canada and 63,000 weekly seats to India [6, 7]. U.S. airlines are restricted from adding any more flights to China until new bilateral negotiations occur [8]. Geopolitical issues mean that these rights, including overfly rights, can be adjusted or even rescinded with little warning.

Just because an airline receives the right to fly to a city does not mean all hurdles are passed. There is a vast seasonality as transatlantic demand is much higher in the summer. Routes to Brazil are packed in January to March, less so in July and August. Some airports are so congested that they are controlled by "slots" or takeoff and landing times.

To address this, the International Air Transport Association (IATA) has a biannual Route Planning and Slot Conference (summer/winter schedule), which determines how frequently and the timing of many international flights [9]. The tricky part is these schedules are set and offered for sale up to 330 days out, and lots can happen in that time to affect the demand.

Further, there may not be an available gate for the desired flight timing, or the airport imposes a "curfew" where no operations occur. This means that airlines may be forced to add flights at undesirable times or not at all. For example, the United States and Japan have "Open Skies." However, Tokyo's Haneda Airport is so congested, the Japanese government restrict the number of flights on U.S. carriers between Haneda and U.S. cities to only four flights per day. Additionally, U.S. carriers were initially restricted to arrival and departure times between 10 p.m. and 7 a.m. [10]. Such restrictions suffocate the market economics for U.S. airlines, even though there is plenty of traffic demand between Tokyo and the United States.

Most carriers treat cargo as secondary, but some flights would not be profitable without the additional revenue from cargo. Specialty all-cargo carriers such as FedEx, UPS, Atlas, and DHL specialize in this market. Several routes haul an impressive amount of cargo. Delta's flight between Detroit and Nagoya, Japan, thrives on automobile industry parts [11]. Flights from South America are often packed with items like Chilean produce, Colombian flowers, and the like [11]. Delta set a record carrying 103,000 pounds of asparagus on a single flight from Lima, Peru [12]. Alaskan salmon season means tons of fresh fish shipped to market in the summer [11]. Significant money is made in the pharmaceutical industry, hauling time- and temperature-sensitive cargo between points [11]. Cargo must be considered carefully when analyzing a route, and the operator must have the appropriate aircraft fleet to support it, along with passenger service.

Fleet Strategy and Products

The type of aircraft to purchase is intertwined with the network decisions, overall business strategies, and of course, cost. Airlines must select a fleet with the cost, size, and range characteristics which will fit its present and future route structure [13]. Southwest Airlines has made a habit of being an all-economy class, Boeing 737 operator. This simplicity keeps costs low, in the form of training, spare parts, and maintenance, while keeping the product consistent [14].

Some civil airline operators serve a mix of short- and longer-range routes, each with varying demand; therefore, they must have appropriately sized and ranged aircraft. Thus, larger network carriers typically have several aircraft types in their fleets. These enable the operators to offer a variety of destinations, capacity, and schedule frequency than if a single aircraft type was flown. Some carriers even have a dedicated fleet of cargo freighters for this market. Indeed, FedEx, UPS, and DHL make good money moving packages, freight, and consumer goods. Increasingly, carriers rely on the "belly capacity" of new-generation aircraft [15].

The cost of an aircraft is not just the purchase price. The operational cost must be considered, often estimated over a 25–30-year life span of operating the asset [16]. Thus, fuel efficiency, engine type, maintenance cost, interior configuration, and inventory/logistic cost are all as important as the capital acquisition cost. Some operators prefer to lease the aircraft for a shorter period of time and periodically renew the fleet. Doing so avoids the capital costs, depreciation costs, and maintenance costs of an aging asset [16].

Large network carriers often have partners, either owned or via contract, that fly smaller-capacity aircraft. This further enables additional smaller cities to be served or additional frequency between some cities [17]. These carriers are often called "regionals," "small-jet providers," or "connection" carriers, but regardless they serve as feeder operations to the larger carriers. An example is SkyWest Airlines in the United States, which provides operations for Delta, United, American, and Alaska Airlines [17]. These smaller carriers serve to "round out" the larger carrier's network.

Often, pilots' unions have "scope clauses" included in their contracts which mandate aircraft above a certain size, weight, or seats must be flown by mainline pilots and not the feeder carriers. Additionally, there are usually caps on the number of aircraft flown by the feeder carriers, with any growth tied to mainline fleet growth.

Airlines also have formed Global Alliances in order to expand their network while managing costs. These are marketed as Oneworld, Skyteam, and Star Alliance and have become commonplace for most large- and medium-sized carriers [17]. Figure 2 shows a Japan Airlines aircraft emblazoned with the Oneworld livery. They operate flights as "code-shares," meaning one airline is able to sell seats on the other airline. This enables larger network opportunities for each carrier. For example, British Airways is able to offer their passengers access to Omaha through partnership with American Airlines, a destination British Airways is unlikely to ever serve itself. Alliances also offer scheduling flexibility, which passengers use as a huge factor in their purchase decision. Delta and

FIGURE 2 Japan Airlines Boeing 777-200ER aircraft painted in the Oneworld livery at Tokyo's Haneda Airport.

Virgin Atlantic combined are able to offer their customers 9 flights per day between New York and London, one of the most important routes in the world, offering competitive frequency when compared with the British Airways and American Airlines alliance, which has 14 daily flights [18].

Joint ventures, protected against antitrust immunity, further enable airlines to share costs and revenues, creating more profit than they each would alone [19]. They have joint marketing and sales, distribution, and often share facilities and employee functions. Some groupings have used the tie-ups to jointly procure items such as office supplies, spare parts, fuel, aircraft, and services at lower cost.

Airlines also compete on product, typically aligned to their business strategy. Emirates Airlines is able to offer such lavish items as an onboard shower, lie-flat seats (converts to a bed), and endless entertainment options [20]. Etihad even offers "The Residence Suite" with personalized butler service in your own three-room apartment [21]. Obviously, these premium products take up more space and are priced much higher but serve as a product differentiator and potential competitive advantage. Other carriers like Ryanair, easyJet, and Southwest offer only coach seating, no meals, and scant amenities.

Airlines can offer an assortment of ancillary products in a bid to entice passengers. Some airline operators offer everything included in the fare: baggage allotment, Wi-Fi access, meals, drinks, entertainment, and even lounge access. Increasingly, however, many operators are moving to the à la carte pricing model where passengers select only the amenities they want. Therefore, these airlines have fees for baggage (both number of pieces and overweight), Wi-Fi, entertainment, and so on. Buy-aboard food and beverage options have become commonplace in the industry, while some operators even charge for seat assignments, upgrades, changes to itineraries, reprinting tickets, lounge access, and even, carry-on baggage.

Finances

Historically, the airlines have been terrible at business. The carriers lost more money between 2001 and 2010 than they had gained from the Wright brothers' first flight until 2001 [17]. Even when they are profitable, the margins are very thin, often in the single digits. This creates issues in the long-term stability of the airline as the carriers are not able to cover the "cost of capital," meaning an inability to pay for fleet renewal and new product offerings [17]. Recently, the larger U.S. carriers have achieved sustained profitability via mergers, lenient bankruptcy laws, and better management decisions than in the past. This has enabled better product offerings, fleet renewals, and other innovations.

Airlines use several metrics to define the finances of a flight. The cost per available seat mile (CASM) and revenue per available seat mile (RASM) are two important metrics [19]. Thus, carriers attempt to maximize the number of seats in a plane, while at the same time they attempt to extract as much money from the passenger as possible. Yield is the term used to describe how much an airline makes per revenue passenger mile and is often measured in mere pennies [19].

Costs

Airlines must control the cost side of the equation very carefully in order to have a chance at profit. A large fixed cost is typically built in before the flight even departs. This includes the cost of the aircraft, facility costs (airport landing fees), product costs, catering costs, and the like. Then, there are variable costs, such as labor and fuel, which are typically the largest costs of an operation [19]. Fuel price is one of those noncontrollable variables, subject to all kinds of impacts. Labor cost is easier to control, as it is often dictated by market rates, but is still tricky. Things like labor work rules, and crew scheduling practices, may impact the ratio of employees to aircraft or flight, thereby affecting cost. Labor unions often present stiff resistance to management proposals which would negatively impact their members. An example is the pilots' unions which include "scope clause" provisions in their contracts.

Productivity is the other aspect of cost. Southwest Airlines traditionally pays their employees the highest wages and is heavily unionized. However, Southwest is also famous for their 20-minute ground times, nearly 30 minutes better than American [17]. This translates into enhanced aircraft utilization, 22% higher than the industry norm, and the ability to offer more flights with fewer aircraft compared to others [17].

Fares and Pricing

Airlines must also determine how to price their products in order to maximize profitability while balancing the time before the product "perishes." They typically have many fare classes from which to choose, depending on the product. The number of seats available in a certain fare category fluctuates based on anticipated demand, schedule (time of year, day of week, time of day), and number of days prior to flight. Each of these also has different rules, like free baggage or not, lounge access, carry-on bag allowances, and checked baggage weight allowance.

Many low-cost carriers offer an à la carte approach to pricing. For example, Ryanair typically offers low basic fares, but then charges for baggage, to print out your ticket at the airport, and a seat assignment, including aisle seat selection as more expensive [22]. The idea is that a passenger can pick and choose the "package" that best suites them. Larger network airlines have recently introduced "Basic Economy" fares to compete with the low-cost carriers, complete with à la carte pricing as well, but offering a wider array of product offerings [23].

The Operation

Running the operation requires a team effort, with a complex variety of skills. If one part of this team doesn't perform well, the success of the operation may be in jeopardy, including logistical, financial, and customer service metrics. A plethora of regulations provide plenty of hurdles as airlines must comply with requirements from regulatory agencies such as the Federal Aviation Administration (FAA), European Aviation Safety Agency (EASA), Transport Canada, the Food and Drug Administration (FDA), Centers for Disease Control (CDC), and the Occupational Safety and Health Administration (OSHA).

Operations Center

Airlines have operations centers which coordinate and oversee all activities of the operation. Because so many aspects make up the operation, the center houses representatives of every work group. These are the people who collectively determine whether to delay or cancel a flight, or proactively make decisions to impact the smallest amount of passengers, all the while keeping everyone safe and complying with all regulations. Housed in a large room, the employees have every bit of information about a flight at their fingertips; see Figure 3.

Reservations/Distribution Channels

Reservationists for the airline perform a variety of customer-facing tasks. They may take phone calls or assist with online chats, and mostly they take reservations. However, other tasks such as re-bookings, upgrades to a different class, baggage questions, and airport facility and lounge questions are also performed. Some airlines have dedicated call centers, and others allow remote reservationists to work from home [24].

Airport Customer Service

Airport customer service personnel assist passengers within the airport environment. They check-in passengers when they first arrive at the airport. They help account for passenger loads, print out tickets, properly tag the baggage, and handle any re-booking questions. Most airlines have gone to online check-in, with the only reason to visit an agent being for baggage drop-off.

At the gates, airport customer service agents help passengers board the plane. They have a large part of the responsibility for having the aircraft depart on time. They also

FIGURE 3 Delta's Operations and Customer Care Center in Atlanta.

serve as one of the few airline employees that passengers actually encounter, so their attitude and disposition are important in this customer-facing role. Roaming agents help with any questions and are especially useful in any irregular operations.

Customer service agents are also found in the lounges and typically cater to premium passengers. These employees are usually specially chosen for their customer service skills as they are significant ambassadors of the airline operator. Technology, such as biometrics, self-tagging of baggage, automated check-in, and automated boarding, is potentially reducing the need for so many employees.

Security

All passengers must go through security checks to gain access to the aircraft. Most security is run by separate governmental organizations, such as the Transportation Security Administration (TSA) in the United States [25]. Security checks include positive identification of passengers, baggage matching, magnetometers (for detection of metal), backscatter x-ray, pat-downs, and even 3D scanners to "digitally unpack" carry-ons [25]. This part of the journey is arguably the most frustrating for the passengers, yet the airlines have little control or input. Technology such as biometrics and facial recognition may lead to improvements.

Aircraft Load Agents

Aircraft load agents, commonly called "baggage handlers," perform the grueling task of loading and unloading passenger bags, plus any cargo. This strenuous job also has the time critical aspect, which is compressed further by the required security check and baggage matching requirements. Additionally, it is an all-weather job, working in the heat of the summer, cold of winter, wind, rain, ice, and snow; see **Figure 4**. While this class of employee does not interact with the customer, they are critical to customer satisfaction metrics such as on-time flights and baggage delivery.

Maintenance

Maintenance is required to run a safe operation. This part of the organization assists with any technical issues and performs required checks to ensure operational reliability. These tasks can be fixing broken seats, changing a tire, adding engine oil, and the like. Line maintenance refers to maintenance that is performed at the gate, while base maintenance or hangar operations perform deeper interrogation of the aircraft, engines, and components. Chapter 8 contains much more information about maintenance and is quite important to operational reliability, on-time measures, delays and cancellations, and, ultimately, passenger satisfaction.

Scheduling

Scheduling is a complex logistical function, which controls the assets of the airline. Typically embedded in the operations center, it encompasses aircraft scheduling, crew scheduling,

FIGURE 4 A rainy day at New York's LaGuardia Airport, and the operation must continue.

and maintenance planning. It involves providing aircraft departures to cities at times (supply) to which and when consumers want to fly (demand) in order to convince them to exchange money (revenue) for those benefits [26]. One source described the task succinctly: "Deciding where airplanes should fly, and at what hours of the day, requires four-dimensional thinking, an affinity for solving puzzles in which the pieces are continually moving… The permutations increase arithmetically according to the number of aircraft, and geometrically according to the number of aircraft types in any given fleet" [4].

Aircraft each have individual itineraries, which can be affected by weather, mechanical issues, air traffic restrictions, or passenger issues such as medical or abusive and uncooperative passengers. Aircraft can be used on as many as 8–10 flights per day, and any delay on one flight can cascade to the others. Additionally, the airport slots, curfews, spare aircraft, and gate assignments must be managed and are difficult even on normal days. Spare aircraft are often kept in reserve in case of mechanical delays of other aircraft. Selecting how many spare aircraft, the aircraft type, and where to position them is a key element in contingent planning of "irregular operations."

Crew scheduling is just as vital as aircraft scheduling. In addition to having the aircraft in the correct location at the right time, the flight attendants and pilots must also be arranged. Additionally, crew scheduling also handles the required hotels and transfers for overnights for the flight attendants and pilots. These must be done under a strict set of regulations, designed to guarantee safety. New regulations (known as FAR 117) mandate daily flight-time limitations, monthly flight time limitations, reserve scheduling, "fitness for duty" guidelines, rest requirements, consecutive nighttime operations, and a new concept called acclimatization based on the theater of operations and other variables [27]. These are tricky to deal with and become a mess when things go wrong.

Maintenance must also be scheduled as required checks must occur at regular intervals. Some can be accomplished overnight while the plane is at the gate. Other checks are longer and require the aircraft to be taken out of service, potentially requiring hangar space and hundreds of man-hours worth of labor. All told, each aircraft receives about 14 hours of maintenance for every hour it flies [19].

Dispatchers

Aircraft "dispatchers" or flight operations officers are 50% responsible (pilots have the other 50%) for the safety of every flight they dispatch. They must have a special license showing their competence in the Federal Aviation Regulations as their function directly affects the safety of the flight [28]. They ensure the aircraft performance, such as proper loading according to maximum takeoff weight, center of gravity, estimated fuel burn, maximum landing weight, and other parameters. They determine how much fuel would be required and how much cargo can be loaded and are embedded in the operations center.

Meteorology

Larger operators have their own meteorology departments, while other carriers contract with meteorology experts. Their function is critical to assist the load planners and the pilots. They must forecast en route storms, icing altitudes, and winds aloft and monitor the weather at both the destination and potential diversion points; see **Figure 5**.

FIGURE 5 Meteorologists are on the staff of many airlines.

Servicing

Servicing aircraft is an important function but is usually outsourced to contractors. These functions include cleaning the aircraft, servicing the lavatories, delivering the catering, fueling the aircraft, and de-icing when required. These are critical to the service delivery and ultimate reputation of the operator (dirty planes, not enough food, overflowing bathrooms, and no drinks). All of the functions must be completed on a rigorous time schedule.

In-Flight Services

In-flight services, or flight attendants, serve as the ultimate customer-facing employee in addition to providing safety in the event of emergencies. Rigorous training is required, including safety, security, medical, cultural awareness, etiquette, customer service, and even self-defense. Additionally, recurrent training is required annually. One flight attendant is required for every 50 seats, and flight attendants usually fly every aircraft type. Close contact with crew scheduling is required in the event of illnesses, reroutes, diversions, trip swaps, and cancellations.

Flight Operations

Flight operations (i.e., pilots) fly the aircraft. They must be licensed to fly and rated on the particular aircraft type. Rigorous annual training and simulator rides are required to prove they can handle any type of emergency. Also, they can be audited at any time by both internal evaluations (i.e., check rides) and by the FAA. They hold the ultimate responsibility of the safety of the flight and must closely coordinate with the operations center, maintenance, airport customer service, and flight attendants.

Corporate Safety and Security

The largest responsibility of an operator is the safety and security of the passengers, employees, and public in general. On-the-job safety is paramount, from simple ergonomics and job hazard assessments to incident analysis and accident reporting to government organizations like OSHA. Emergency evacuation plans, chemical/fuel spill planning, electrical safety guidance, and other items are examined for risks.

Consultations are made with law enforcement officers, plus national security levels of government and the CDC. This can be researching terrorism risks, epidemic and disease precautions, government protests, plus any natural disasters. Each city is examined for these security risks, both at the airport and the layover hotels, and daily updates are provided to management and the operations center for disposition. These typically are "cautions" only but can include cancelling flights and even market suspensions.

Other Supporting Functions

Each airline also has plenty of behind-the-scenes people working administrative functions. Subjects like accounting and finance, Information Technology (IT), marketing, sales, engineering, and general administration are all required functions. IT is extremely important in this digital world from back-office items like scheduling programs, dispatch programs, computer safety, corporate records, and privacy laws. Customer-facing IT is probably more important with kiosks, baggage tagging, and mobile apps.

When Things Go Wrong

Despite the complex logistics of the operation, airline operators get it right the vast majority of the time. The flights depart and arrive on time, people claim their baggage, and passengers are mostly satisfied. However things do go wrong and are termed "irregular operations." Weather impacts the operation, baggage is lost or damaged, irate or inebriated passengers might be disruptive, incomplete catering, and a host of problems all provide hurdles for the airline operator.

When weather forces planes to divert (e.g., thunderstorms, hurricanes, icing conditions), the airline has passengers, airplanes, flight attendants, and pilots in the wrong places. The impact on passengers is obvious, while the pilots and flight attendants are subject to the various regulations and rules of duty time, minimum rest periods, etc. Thus, it is common to hear a diverted flight impacted by the crew "timing out," potentially driving a cancellation. Another aircraft, typically a spare, and crew are flown in to recover the passengers on that particular flight. This is where "reserve" crews are utilized.

The impact however has a cascading effect, not just for one particular flight. An instantaneous recovery is often not possible for a major irregular operation, as planes and crews must be moved to catch up and resume the schedule. A simple thunderstorm may set back the operation by hours, while larger events (hurricanes, for instance) take days to recover. Passengers may be baffled how a storm two days prior has affected "crew rest," delaying their flight. Or how weather elsewhere earlier in the day impacts their current flight, as the aircraft due to deliver their flight was delayed four flights ago. All in all, the ability and effectiveness of an airline operator to recover from these events can distinguish itself from their peers [29].

The Human Part: Employees

Some of the trickiest parts of the operation are human related. Airports abound with a variety of airline employees and contracted providers. Each must carefully be managed within the operation, which runs 24/7, 365 days a year. Labor problems have been commonplace in the industry. Strikes by any one group, including groups outside the airline such as air traffic controllers, can grind the operation to a halt.

Flight personnel with whom passengers will come into contact are largely unsupervised. The fundamental challenge of airline management is to motivate these large numbers of somewhat inherently discontented and unsupervised employees to exude courtesy and friendliness to the airline's customers [30]. For years, Southwest has led the industry in customer service (i.e., fewest customer complaints), despite offering scant amenities. To constantly remind employees, "each payroll check is inscribed with the words "From our Customers" to remind the employee who is buttering their bread [4].

The relationship between management and airline employees is key, and if it becomes too toxic, it may cause the downfall of the airline, as experienced by Eastern [4].

Outsourcing of work has been a trend in attempt to lower labor costs. It has been joked that airlines would one day have only one employee, the boss, with every other job function contracted or outsourced [31]. Indeed, some carriers have resorted to temporary "contracts" for pilots and flight attendants, with renewal at the airline's option [31]. Labor unions fight for jobs, pay rates, and benefits for their members but typically must compete against the outsourcing trend. Maintenance costs can be lowered by sending work to maintenance, repair, and overhauls (MROs), often in different countries where labor rates are lower, like Eastern Europe, Mexico, and China.

In cities where airlines have small operations, they typically outsource functions in order to be more cost efficient. It makes no sense for an airline to pay employees full time for a daily operation of a single flight. These include baggage handlers, ticketing agents, and gate agents. However, Alaska Airlines changed the landscape when they outsourced baggage operations in their hub of Seattle to Menzies [32]. The outsourcing could have an impact on customer service and product quality.

A number of restrictions shape the labor landscape. Maximum duty times (daily and monthly) and minimum crew rest drive the schedule for pilots and flight attendants. Likewise, vacation time, sick days, and overtime rules all impact the operation, both from a cost standpoint and from a logistical standpoint. New regulations (known as FAR 117) mandate daily flight time limitations, "fitness for duty" guidelines, rest requirements, consecutive nighttime operations, and a new concept called acclimatization based on the theater of operations and other variables [27]. These rules must be complied with, including during events like diversions, delays, and cancellations, which add complexity to the operational scheduling.

Seniority governs most employee groups at the airline. This could be which shift they work, what off-days they receive, order of preference for vacations, and in the case of pilots and flight attendants where and when they fly (and whether they layover in Paris, France, or Lubbock, TX) [30]. Many employees work odd hours, going against their circadian rhythms by crossing multiple time zones (pilots, flight attendants) or working overnight (maintenance). For every 1-hour time zone difference, the body requires 1 day to adjust its biological clock [30]. Pilots and flight attendants are often away from their families for days at a time, and holidays must be worked. Some even "commute" by living in a different city than they are based. These human factors must be considered as the employees are the ones who deliver the service to the passengers.

When the operation is impacted, employees are directly impacted. This is where the operations center directs the operation around the various hurdles, including crew scheduling, aircraft scheduling, and meteorology. Flight attendants and pilots are rerouted, spending the night in Cleveland instead of Miami as planned. Other employee groups such as airport customer services, aircraft loading agents, and maintenance may be required to work overtime in order to get things back on track.

Proactive planning from the operations center can help ensure the impact to the operation is minimized. This may be pre-positioning spare aircraft to a particular city in advance of a storm, rerouting passengers through alternate hubs, or proactively cancelling a flight early in order to avoid a diversion or passengers being stranded at the airport.

The Human Part: Passengers

Passengers must cooperate through the various machinations of check-in, security, getting to the gate, and boarding. Then they are squeezed into a small tube with a few hundred strangers for hours on end. It is little surprise that there are reports of upset passengers assaulting flight attendants or fights between passengers. Add delays, cancellations, involuntary bumping, and alcohol to the mix, and one has a potentially toxic situation.

Flying used to be considered a luxury, but now is commonplace among developed areas of the world. As air travel has become more affordable, the masses can now travel, bringing full flights and crowded airports. Passengers are told to stand in one line (check-in), then must go to another line (baggage drop), then another (security), and fight the mass of humanity to be able to stand in another line (boarding). Upon arrival, they reverse the process by standing in line for deplaning, immigration, baggage retrieval, and customs. **Figure 6** shows crowds of people waiting for immigration and passport control at Singapore's Changi Airport.

Airlines continually squeeze more seats into planes, reducing personal space for individual passengers. Some passenger rights groups have complained that airlines treat their passengers like "cattle," herding them from on point to another [33]. When things do go wrong, some airlines treat passengers even worse, as documented on social media [34].

FIGURE 6 Crowds of people await passport control at Singapore's Changi Airport.

Humans are a flexible bunch, able to withstand a variety of occurrences. However, when they are not fed, get little sleep, drink too much, are claustrophobic, or have a fear of flying, their patience may wear thin.

When they feel disrespected by not having an update of a delay, fighting for elbow room, or receive poor treatment, they may explode with anger.

Passengers also travel for a plethora of reasons, which may add to the stress level of an individual. The passenger may be flying to a funeral, a potential career-making deal, or other stressful event. When delays or cancellations occur, passengers may miss their daughter's recital, dream vacation with their spouse, or the big meeting with a potential client. All this adds to the anger when an airline does not deliver the expected service.

Many consumer regulations have been forced upon the airlines, primarily due to blatant mistreatment of passengers, intended or otherwise. The most stringent of these is in Europe, where Regulation 261/2004 detailed the requirements for compensation of delayed and cancelled flights [35]. In the United States, legislation was enacted to require access to food and water when delayed on the tarmac. Airlines can be fined for tarmac delays over 3 hours for domestic flights and 4 hours for international flights [36]. The United States also requires airlines and airports to comply with the Americans with Disabilities Act, which assures access to people with disabilities (e.g., wheelchair ramps instead of stairs, special accommodations on-board, and larger bathrooms) [37].

The U.S. Department of Transportation also tracks many metrics which grade the airlines monthly in on-time performance, completion factor, percent of lost/delayed baggage, and denied passenger boarding [38]. These rankings are very competitive for the airlines, as it could impact future bookings if an airline is consistently a poor performer. Many consumer groups continue to press for passenger-friendly practices, either voluntarily by the airlines or forced via legislation. Hot topics include minimum seat width, minimum legroom, endless fees, and fare caps via pricing regulation [39]. It is anticipated that legislation similar to Europe Regulation 261 for compensation of delayed and cancelled flights will spread throughout the developed world.

The airline operation is complex, and operators perform excellently most of the time, as many people forget. Instead, airlines are often judged by how well they perform during irregular operations, as documented by instant social media outlets. It is up to the airlines to ensure passengers are treated correctly, even when delays and cancellations occur. Passengers must also behave in a civilized manner and exercise patience when things go wrong.

References

1. American Customer Satisfaction Index (ACSI), 2018, accessed January 16, 2018, http://www.theacsi.org/index.php?option=com_content&view=article&id=147&catid=&Itemid=212&i=all&sort=Y2015.
2. FAA, 2018, accessed January 2, 2018, https://www.faa.gov/regulations_policies/.com.
3. Delta Flight Museum, 2018, accessed January 15, 2018, https://www.deltamuseum.org/exhibits/delta-history/first-in-the-air.

4. Petzinger, T., *Hard Landing* (New York: Crown Business Publishing, 1995), ISBN 978-08129238358.
5. Bureau of Transportation Statistics, *Airlines and Airports* (Washington: U.S. Department of Transportation, 2017).
6. Chowdhury, A., "Dubai Seeking to Expand Air Traffic with India," *The Economic Times*, February 16, 2016.
7. Jang, B., "Airlines Battle 'One-Sided' Agreement with Emirates," *The Globe and Mail*, October 13, 2010.
8. Yan, F., "US-China Aviation Talks Hit Stumbling Block on Airport Access," Reuters, August 13, 2015.
9. International Air Transport Association (IATA), *Worldwide Slot Guidelines*, 8th ed. (Canada: IATA, 2017).
10. Kageyama, Y., "U.S. Carriers Feel Squeeze at Tokyo's Haneda Airport," Honolulu Star Advertiser, September 13, 2010.
11. Morrell, P., *Moving Boxes by Air: The Economics of International Air Cargo* (UK: Routledge, 2016), ISBN 97813170993824.
12. Delta, "Delta Increases Capacity between Lima and Atlanta," Delta Press Release, March 30, 2016.
13. O'Conner, W., *An Introduction to Airline Economic*, 6th ed. (Westport: Praeger Publishers, 2000), ISBN 0-275-94863-3.
14. Gulliver, "The Secrets of Southwest's Continued Success," *The Economist*, June 18, 2012.
15. Johnsson, J., "Big-Belly Boeing 777 Blunts Demand for Cargo-Only Jets," Bloomberg, June 26, 2014.
16. Flottau, J., "Person of the Year: Delta Air Lines' Richard Anderson," *Aviation Week & Space Technology*, December 18, 2015.
17. Dempsey, P., *Airline Management: Strategies for the 21st Century*, 3rd ed. (Chandler: Coast-Aire Publications, 2010), ISBN 978-1890938093.
18. Delta, "Delta and Virgin Update Network between U.S. and U.K.," Delta Press Release, September 26, 2017.
19. Jenkins, D., *Handbook of Airline Economics*, 2nd ed. (New York: Aviation Week Publishers, 2002), ISBN 978-0079823946.
20. Emirates, 2018, accessed January 13, 2018, https://emirates.com.
21. Etihad, 2018, accessed January 13, 2018, https://etihad.com.
22. Ryanair, 2018, accessed January 15, 2018, https://ryanair.com.
23. McGee, W., "What You Should Know about Basic Economy Airfares," Consumer Reports, August 9, 2017.
24. WAH, "These 27 Companies Will Hire You to Work from Home Taking Reservations," 2017, assessed January 15, 2018, https://wahadventures.com.
25. TSA, "Packing for Your Next Flight," 2018, accessed January 15, 2018, https://tsa.gov.
26. Banfe, C., *Airline Management* (USA: Prentice Hall, 1992), ISBN 978-0130191830.
27. FAA, 2013, accessed January 16, 2018, https://far117understanding.com/understanding-the-new-regulation/basic-concepts-of-far-part-117.
28. Houston, S., "How to Become and Aircraft Dispatcher," *The Balance*, October 22, 2017.
29. Transportation Research Board. *Guidelines for Airport Irregular Operations (IROPS) Contingency Planning* (Washington: TRB, 2016).

30. Dempsey, P., *Air Transportation: Foundations for the 21st Century*, 2nd ed. (Chandler: Coast-Aire Publications, 1997), ISBN 978-1890939123.
31. Wild, J. "Airlines Turn to Outsourcing to Keep Lid on Costs," *Financial Times*, April 21, 2014.
32. Allison, M., "Alaska Airlines Outsources 472 Baggage-Handling Jobs," *Seattle Times*, May 14, 2005.
33. Josephs, L., "Warren Buffett: Airlines are Becoming "Cattle Cars" Because That's What They Want," *Quartz Magazine*, May 8, 2017.
34. Mills, K.A., "Dr. David Dao's Lawyer Hammers United Airlines Arguing Passengers 'Should Not Be Treated Like Cattle'" *Daily Mirror*, April 14, 2017.
35. Europa, 2004, accessed June 14, 2017, http://eur-lex.europa.eu/resource.html?uri=cellar:439cd3a7-fd3c-4da7-8bf4-b0f60600c1d6.0004.02/DOC_1&format=PDF.
36. U.S. DOT, "New DOT Rule Consumer Rule Limits Airline Tarmac Delays," 2018, accessed January 16, 2018, https:www./transportation.gov.
37. U.S. DOT, "Passengers with Disabilities," 2018, accessed January 16, 2018, https:www./transportation.gov.
38. U.S. DOT, "Air Travel Consumer Reports," 2018, accessed January 16, 2018, https:www./transportation.gov.
39. Wattles, J., "Judges Order FAA to Review Airplane Seat Sizes," *CNN*, July 29, 2017.

CHAPTER 8

Maintenance: A Necessary Evil?

David Piotrowski
Delta Air Lines

God created aircraft mechanics, so pilots can have heroes too.

Anonymous

Introduction

Safety and compliance are the most important things to consider when it comes to aviation maintenance and always will be. In this highly regulated industry, airlines and maintenance, repair, and overhaul (MRO) organizations carry out original equipment manufacturer (OEM) originated, scheduled-based programs to move passengers and cargo safely. Airlines compete against each other for schedule performance (on-time percentage, completion factor), baggage delivery, and amount of cancellations while moving passengers and cargo. With aviation maintenance so regulated, one could presume all carriers perform maintenance the same. However, a closer look reveals many interesting and complex nuances within the aviation maintenance realm.

Maintenance has often been described in the past as a necessary evil in asset-rich organizations, typically perpetuated from the financial side of that organization [1]. Airlines do not make money if an aircraft is on the ground, broken, and unscheduled out of service. Increasingly, the cost of lost revenue dwarfs the cost required to fix the asset.

Additionally, passengers and cargo are negatively impacted by these operational disruptions, such as cancelled, delayed, or diverted flights, which then require restitution from the airline. In February 2004, the European Parliament and Council of 11 established "common rules of compensation and assistance to passengers in the event of denied boarding and cancellation or long delay of flights," and the resulting Regulation 261/2004 detailed the requirements [2]. In addition to the negative financial impact of an operational disruption, there is an unquantifiable impact to the brand or reputation, which could affect future earnings for the airline. Therefore, good maintenance organizations or effective MROs are now highly valued and indeed even a competitive advantage for some airlines.

Business models and relationships are complex and evolving, with even OEMs offering MRO services. Third parties often have innovative solutions or products, but there are big barriers to market entry. Airlines are diverging on maintenance philosophy—some build up MROs, and some outsource their maintenance. Thus, some airlines are customers and competitors to OEMs simultaneously.

This realization of how important maintenance is to the bottom line of an airline has provided the impetus to find every way possible to avoid an operational disruption. A modern rationale for integrated vehicle health management (IVHM) comes from a transformation that OEMs have gone through, from selling a product in which future income is dependent on spare parts revenue to selling a service in which steady monthly income can be derived in return for effective maintenance of the asset [1]. Most OEMs, and even some MROs, now offer some type of this service to the airlines. With "Big Data" and analytics, this service niche has tremendous growth potential.

It is noted that the author works for Delta Air Lines, so most examples in this chapter come from the author's experiences while working at Delta. These are not intended to be commercial advertisements but serve as examples of innovation occurring in the industry.

Safety and Compliance

The civil aviation maintenance industry must be safe for passengers and employees and compliant with rules and regulations. The smallest slip can be fatal, causing loss of life. In the early days of civil aviation, maintenance reliability was poor, leading to many accidents and disrupted operations. Thankfully, improper maintenance is rarely the cause of airline accidents these days. This is not accomplished by luck. Instead, a rigorous set of laws, standards, and procedures are followed, with plenty of auditing and double-checking.

The International Civil Aviation Organization (ICAO) has published in their Safety Management Manual, a "safety cycle" [3]. This is shown schematically in **Figure 1** and includes both the design stage and operational stage. In this section we focus on the operational stage; for the design stage details, see Chapter 11. The operational "safety cycle" identifies the hazard, assesses the risk, evaluates options, and then communicates and directs maintenance action (i.e., replace the part, conduct inspection, etc.) designed to mitigate the hazard. The final step is to monitor the progress and then continuously repeat the cycle. The steps are formalized in the regulations that the operators must follow.

FIGURE 1 In the ICAO "safety cycle," the safety is assured by the maintenance operation.

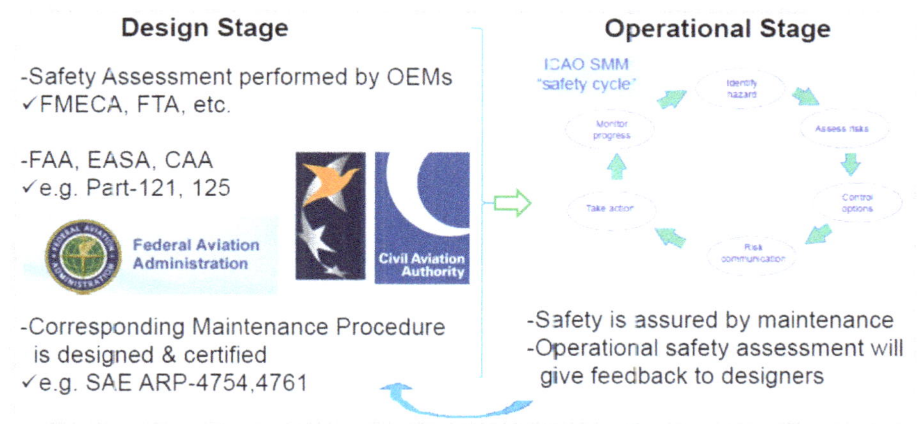

Regulations

Airline maintenance is highly regulated because of the significant consequence of errors. The ICAO sets global standards which are then implemented and enforced by national or regional bodies [4]. These include Agência Nacional de Aviação Civil (ANAC) in Brazil, Civil Aviation Administration of China (CAAC) in China, Civil Aviation Authority (CAA) in the United Kingdom, Civil Aviation Safety Authority (CASA) in Australia, Directorate General of Civil Aviation (DGCA) in India, European Aviation Safety Agency (EASA) for Europe, Federal Aviation Administration (FAA) for the United States, and Transport Canada (TC) in Canada, among others.

In the United States, Title 14 of the Code of Federal Regulations (CFRs) and Federal Aviation Regulations (FARs) dictate the rules governing all aviation activities in the United States [5]. These FARs are divided into parts, with Part 21 (Certification Procedures for Products and Parts), Part 25 (Airworthiness Standards: Transport Category Airplanes), Part 43 (Maintenance, Preventive Maintenance, Rebuilding, and Alteration), Part 91 (General Operating and Flight Rules), Part 121 (Operating Requirements: Domestic, Flag, and Supplemental Operations), and Part 125 (Certification and Operations: Airplanes Having a Seating Capacity of 20 or More Passengers or a Payload Capacity of 6,000 Pounds or More) relating to airline maintenance. Smaller operators also reference Part 135 (Operating Requirements: Commuter and On-Demand Operations), while MROs are also governed by Part 145 (Repair Stations).

Maintenance Programs and Planning

The current maintenance philosophy is based on the MSG-3 (Maintenance Steering Group-3) document, which established a disciplined decision logic and procedures for determining maintenance plans. This includes an Industry Steering Group which consists of the OEMs, operators, and regulators. It provides a unified maintenance plan, which the FAA documents in a Maintenance Review Board (MRB) report. The primary purpose of this report is to determine the initial scheduled maintenance requirements for new or derivative aircraft. The OEMs then produce a Maintenance Planning Document

FIGURE 2 A mechanic is checking tire pressure, evidence of tire wear, and general condition of the landing gear and brakes.

(MPD) for individual tasks which meet the requirements set forth in the MRB. It includes maintenance labor hours, facilities and tooling recommendations, and recommendations on task packaging strategies [4]. It is a living document which is constantly updated and is different for each aircraft type and/or usage. Job cards then detail the individual tasks to be accomplished, whether routine or nonroutine.

Planning to accomplish these tasks leads to a series of checks, based on various flight cycles, flight hours, or calendar days. Service checks are the lowest level of scheduled check, typically conducted every few days. Examples of tasks under a service check are checking fluid levels and tire pressure and wear, ensuring emergency equipment is installed and serviceable, and general security and cleanliness [4]. Figure 2 shows an example of a service check, with a mechanic checking for tire wear.

Letter checks, named A through D, then begin to examine the airplane and engines with more detailed inspections, requiring more time, special equipment, special tooling, and personnel. The A-check is often conducted every few weeks, typically overnight, and consists of tasks like landing gear lubrication, oxygen system pressure check, hydraulic pressure check, and a general external visual inspection for evidence of damage, deformation, corrosion, cracking, or missing parts. The B-check is a slightly more detailed check of components and systems and occurs every few months. The C-check, shown in Figure 3, occurs roughly every 1.5–2 years in the hangar. It involves more invasive checks, including all the tasks from service checks, A-checks, and B-checks. Examples of these types of tasks include visual inspection of door and window seals, operational checks of flight systems (flaps, elevators), and some structure checks for cracking, corrosion, or delamination in the case of composites [4]. The D-check is the most invasive and detailed check and often requires the complete removal of the interior down to the "skin and bones" of the aircraft; see Figure 4. These typically occur every ~6 years or so and are heavily focused on structural inspections.

FIGURE 3 An aircraft in the hangar, undergoing work as part of a C-check.

FIGURE 4 An aircraft undergoing a heavy "D-check," which strips the aircraft down to the structure.

There are other nuances of scheduled aircraft maintenance, often specifically designed to support operator preferences. Some airlines use phased checks, which divide the C- and D-checks into segmented blocks of maintenance tasks. These segmented blocks can be distributed to the more frequent, but shorter, checks. This avoids taking an aircraft out of service for the extended periods of time associated with C- and D-checks. Some D-checks can be as long as 50–60 days. Some smaller airlines even divide the items into packages which can be accomplished from night to night.

Specialized maintenance and inspection programs are required additions for certain operations. Extended twin-engine operations (ETOPS) contain even more regulations to produce even higher reliability on long flights due to the large distances to alternate or diversion airports from the flight path. Maintenance-wise, these include specialized navigation equipment checks, and more frequent checks of fluids, general condition, and security. These checks often are required before every flight.

The OEMs must support their products while in operation. Often, they will provide service bulletins (SB) which supply guidance to the operators on particular items such as design improvements or new recommendation which may affect operational schedule (i.e., delays and cancellations). These can encompass recommended part replacement, implementation of additional inspection tasks, specific repair procedures, and the like. There are several different types of SB. Most are typically not required but are recommended or optional actions. Alert SB address a condition the manufacturer feels is a safety-related item as opposed to just a product improvement. These usually result in the FAA issuing an airworthiness directive (AD). ADs are legally enforceable rules issued by the FAA to correct an unsafe condition in an aircraft, engine, propeller, or appliance [6]. These require the airlines and MROs to comply with specific maintenance actions within a specified timeframe. ADs occasionally mandate the accomplishment of a SB, which may have been optional prior to the AD release.

Aging aircraft also require more frequent maintenance tasks, such as additional supplemental structural inspections. These special programs are required after the aircraft accumulates a large amount of aircraft cycles or hours and are typically for detection of fatigue cracking. New rules for damage tolerance analysis and prevention of widespread fatigue damage (WFD) further added maintenance requirements. Some programs are called supplemental structural inspections (SSIs), and others are called airworthiness limitation items (ALIs). The operators also must have corrosion prevention and control programs (CPCP). These inspect for corrosion, repair corrosion-affected areas, apply corrosion inhibiting compounds, and require reporting to the FAA. Additionally, the required implementation of the CPCP, SSI, and ALI programs noted above are all mandated by AD.

Maintenance Evolution

Maintenance has experienced an evolution from "run to failure" modes of operation, through time-based solutions such as scheduled maintenance, preventative maintenance, and reliability-centered maintenance (RCM), to real-time product performance measures including new philosophies like condition-based maintenance (CBM), engine health management, system health management, and structural health management [1, 7, 8].

In the late 1960s, civil aviation maintenance saw a transformation via the Maintenance Steering Group (MSG) process, initially for the Boeing 747. This new maintenance program was determined to be operationally and economically viable, which was a concern of the industry. In 1978, the Department of Defense commissioned the infamous RCM report [8], leading the civil aviation sector to upgrade to MSG-3. Since then, the civil aviation industry has seen monumental improvements in both safety and reliability. The industry is now undergoing further evolution with the addition of IVHM systems combined with "Big Data" to enable more prognostic and predictive maintenance.

Aircraft health monitoring (AHM) is now taking the industry to the next iteration of maintenance programs. AHM serves as an umbrella which contains engine health monitoring, component health monitoring, systems health monitoring, and structural health monitoring (SHM). Delta worked with Boeing to provide an alternate inspection program using SHM on Boeing 737s [9]. The program also served as a "cradle-to-grave" program involving the FAA, which was interested if any further guidance or policy was required. It examined all aspects of SHM including certification, validation, installation, usage, and monitoring for the operator, the OEM, MROs, regulators, and technology providers [9].

The industry must navigate these evolutions while remaining compliant with stringent regulations. In fact, civil aviation also has more rules and laws (FAA, OSHA, FDA) than almost any other industry. Ultimately, some industry documents must be revised, or newly created, to pave the path for the renaissance of these new maintenance services, whether from the OEMs, MROs, operators, or third parties.

Big Barriers to Entry

Clearly, there is money to be made in maintenance services. Airliner MRO reached $74.3 billion in 2017: 51% ($37.9B) single-aisles, 21% ($15.6B) long range twin-aisles, 8% ($5.9B) medium range twin-aisles, 7% ($5.2B) large aircraft, 6% ($4.5B) regional jets as turboprop regional airliners, and 1% ($0.7B) short range twin-aisles [10]. However, in order to penetrate this maintenance service niche, big barriers must be overcome. Mostly this comes from significant regulatory requirements but also includes financial, reliability, and intellectual property implications.

Manuals, procedures, drawings, and other data are required for the maintenance activities. These are usually produced by the OEMs for the operators. Without this data, the CFRs cannot easily be met, and businesses cannot sell their product or service. Increasingly, OEMs are safeguarding their intellectual property, only making it available for significant cost, or outright refusing to provide it to potential competition. MROs unaffiliated with airline operators are especially hamstrung since OEMs usually only provide the information to operators.

If a company were to provide a service, the reliability requirements are extremely high. For example, the reliability of an IVHM system must be higher than the asset it is attached to. This is not a simple task in the real-world environment of pressurization cycles and loads, temperature cycling, exposure to many fluids, greases, and gases, environmental conditions, weather, and so on. Additionally, there are often significant financial penalties written into service contracts for missing reliability targets, spare parts availability, and turn time.

Business Models Are Complex

It has often been said that the OEMs know how to build aircraft, engines, and components, but it is the airlines and MROs who know how to operate and maintain those assets. Additionally, the competition for new aircraft and engine sales is very intense, often leading to significant pricing discounts in order to win the order. Thus, the OEMs have tended to make their money in spare parts sales [11].

Business models are a lot more complex than selling a product and hoping to sell spare parts. Due to low-cost component suppliers and Parts Manufacture Authority (PMA) providers, the OEMs have experienced an erosion in percentage of spare parts revenue. OEMs looking to increase or maintain revenue have moved into maintenance services, often under an hourly contract, a fee for each flight, or monthly subscription fees. In such an arrangement, the OEM is able to use its product knowledge to manage the operation, risk, spares, and so on and acquire more information about the product's through-life performance [1, 12]. Additionally, by offering such a service, the OEM may be able to protect the spare parts business from competition. Benedettini et al. introduced the usage of IVHM as a paradigm shift in the strategy of organizations across all industrial sectors, possessing the ability to change the entire business model via the use of prognostics [13].

OEMs Expanding into MRO Services

Most OEMs now offer a suite of service options to the operators. Some are merely spare parts contracts, guaranteeing availability, turn time, and pricing. However, in order to provide these services, the OEMs must have access to the airline operators' service usage history, operating conditions, failure data, and reliability metrics. Some airlines consider this information to be proprietary. Therefore, the OEMs offer various suites of options to tailor to the needs and desire of the airline or MRO.

Rolls-Royce now offers CareServices—including TotalCare® Life®, TotalCare® Term®, TotalCare® Flex®, SelectCare™, LessorCare, and Foundation Services. These comprise various core and optional services, designed to allow the operator to select the scope of service solutions. Rolls-Royce also pioneered the "Power-by-the-Hour™" concept in 1962 [14]. Rather than simply maintenance, options include other areas such as availability, efficiency, asset management, transitions, and customer support [15]. GE also offers four options for aftermarket services: TrueChoice™ Flight Hour, TrueChoice™ Overhaul, TrueChoice™ Material, and TrueChoice™ Transitions [16]. Boeing has Global Fleet Care, Pratt & Whitney has EngineWise™, and UTC Aerospace Systems, Honeywell, and Airbus also offer such services [17, 18]. By offering these services, the OEMs have secured a path for acquiring the information needed from the airlines and MROs.

OEMs are pushing into the growing service sector due to the aftermarket's stability compared to the cyclic nature of aircraft orders. Whereas Boeing and Airbus primarily have a duopoly on commercial aircraft orders, the aftermarket service market is fragmented. Airbus expects the global demand for new civil aircraft from 2017 to 2036 will be $5.3 trillion, compared to $3.2 trillion for global services. Therefore, this is one of the reasons for massive consolidation in the market such as Safran, Zodiac, UTC, and Rockwell Collins, with more likely to come [19].

Some OEMs have either purchased other companies or partnered with key airlines and MROs to build a service network. Boeing bought Aviall, Inc. to provide airline customers aviation parts and services [20]. Rockwell Collins bought BAE Aerospace but then was bought by UTC [21]. Safran and Zodiac are combining into an aircraft systems giant [21]. GE and Pratt & Whitney have their own overhaul shops around the world. Rolls-Royce has wholly owned MRO facilities, joint ventures, and has partnered with "independently owned" airline or MRO facilities to provide an authorized maintenance and service center network. Delta TechOps is one of those partners with Rolls-Royce [22].

Airlines and MROs

Airlines also vary in their philosophies about MRO. Some airlines, such as Delta, Lufthansa, Air France-KLM, and British Airways, have such economies of scale they prefer to handle most everything in-house. Other airlines may only have a handful of aircraft and elect to outsource their maintenance. It is these smaller carriers which have been most open to outsourcing services. However, the larger airlines must have spare parts and often negotiate volume-discounted deals with the OEMs. This provides the impetus for the OEMs to offer such a wide array of aftermarket service options.

Some airlines have leveraged their expertise and created MROs which offer services to other airlines. Delta TechOps serves more than 150 aviation and airline customers from around the world, specializing in high-skill work such as engines, components, avionics, airframe, and line maintenance [22]. Delta has a maintenance cost per available seat mile (ASM) at $1.21 cents versus the industry average of $1.35 cents [23], providing a great reason for other airlines to want their maintenance to be performed by Delta. Lufthansa Technik has built one of the world's largest networks of MRO services with more than 30 subsidiaries and affiliates, bringing in over 5 billion euros in annual revenue [24]. The insourcing of profitable maintenance work has helped contribute to the bottom line of the parent carrier and even smoothed out some of the cyclic nature of the aviation industry.

Reducing Costs with Innovation

This has provided an interesting industry dynamic where some airlines are both customers and competition for the OEMs. Controversies have erupted over pricing, PMAs, liability, intellectual property, and parts availability. However, cost pressures and access to data are forcing airlines, MROs, and OEMs to find innovative ways to cooperate. This has brought about the "authorized service centers," joint ventures, and other business tie-ups.

The aviation industry is a difficult one in which to innovate. Other industries find it much easier to incorporate new ideas, processes, products, and services. This is primarily due to the stringent regulations which govern the industry under the mantra of safety and compliance. In order to prove compliance to the CFRs, new product or service must pass rigorous testing such as flammability, aerodynamic, environmental, durability, acoustical, and electromagnetic interference and have Instructions for Continued Airworthiness. Additionally, software and IT verifications must occur for cybersecurity, scalability, end-user testing, and Service Level Assurance. Thus, airlines are only now rolling out Wi-Fi onboard aircraft. Electronic Flight Bags, containing all required information for pilots, have only recently replaced paper manuals, maps, and magnetic compasses in the cockpit. Third-party companies are at a significant disadvantage since certification often requires data which is inaccessible to them. Instead, they often must partner with the OEMs in order to succeed. APB Winglets Co. is a prime example where companies providing the technology existed but didn't fully succeed until partnering with an OEM [25].

Health Monitoring, Big Data, and Analytics

Thus, OEMs are in prime position to lead the introduction of innovative products and services. Providing an entire suite of offerings for maintenance services and establishing MRO networks have provided the OEMs with data required to launch health monitoring services. Boeing offers their Airplane Health Management (AHM), while Airbus offers a Prognostics and Risk Management solution, which are real-time decision support tools that allow for faster, data-driven decisions by converting airplane data into actionable information [17, 26]. This results in a reduction in unscheduled maintenance and increased aircraft availability (i.e., fewer delays and cancellations) and enables predictive maintenance prognostic tools. Airbus has revealed that up to 15% of A330 operational disruptions could be avoided using such a program [27].

The airlines are enjoying the benefits of the OEM solutions but are also generating innovative products and services themselves. Lufthansa Technik received a patent for a "System and Method for Monitoring a Nickel Cadmium Battery in a Passenger Aircraft" [24]. Delta has patent pending on a few items, including a monitoring system for the B737NG environmental control system [28].

The field of health monitoring is leading to more and more data being acquired, analyzed, and monitored. Instead of being required to perform scheduled maintenance at particular intervals, the technology is allowing for data-driven maintenance practices. These "maintenance credits" will provide the next stage of evolution of the airline maintenance program. This will accelerate in the future as "Big Data" and better analytics will provide more predictive maintenance solutions, leading to improved reliability and significant cost savings. An early example is Delta, which is using data analytics to differentiate its operational performance and has avoided over 1,000 potential negative engine events over a 12-month period in 2016–2017 [19].

RFID

The industry is also beginning to leverage Radio Frequency Identification (RFID) technology for cost savings, configuration control, part tracking, shelf-life verification, and increased compliance integrity. Delta TechOps, United, Qantas, British Airways, Air New Zealand, Fiji Airways, Swiss Airlines, easyJet, Norwegian Airlines, and others have utilized RFID to track life vests, emergency equipment, rafts, slides, and oxygen generators on and off the aircraft [29, 30]. These are items which require frequent visual checks of their presence on the aircraft, their part number configuration, and their expiration date. Instead of having a group of mechanics take several hours to examine and document each item, a simple walk-through the cabin with a handheld device accomplishes the task. Consequently, if an item requires replacement, it can be located quickly, saving further time. This technology is expanding rapidly to other areas, including tracking passenger bags and cargo.

STCs

A common path for new products and services is via a Supplemental Type Certificate (STC). This is issued by the regulatory agency (FAA, EASA, DGAC, CAA, ANAC) for a major design change when the change is not so extensive so as to require a new type certificate. They are usually very restrictive, limiting the design change to a specific range of airplane serial numbers rather than the complete model series. Examples are the installation of Wi-Fi systems, new overhead bins, or repair process.

Delta has partnered with many companies to earn STCs. Delta worked with Lambda Technologies to create a method for using Low Plasticity Burnishing (LPB®), to mitigate fretting-induced fatigue cracking in the high-pressure compressor section of the CFM56-7 [31]. The technology was shown to increase the fatigue life of a damaged surface more than 10 times. This results in better reliability but also allows Delta to reduce the cost of spare parts.

Lufthansa Technik was recently approved for an STC after they developed a new antenna radome for faster data transfer [24]. This provides higher speed rates for Internet and TV connections, as well as enabling a conduit for real-time health monitoring. Additionally, the system saves weight and was shown to reduce aerodynamic drag, providing cost savings.

References

1. Jennions, I.. 2011. *Integrated Vehicle Health Management, Perspectives on an Emerging Field*. SAE International. Warrendale. ISBN 978-0-7680-6432-2.
2. Europa, 2004, accessed June 14, 2017, Website: http://eur-lex.europa.eu/resource.html?uri=cellar:439cd3a7-fd3c-4da7-8bf4-b0f60600c1d6.0004.02/DOC_1&format=PDF.
3. International Civil Aviation Organization (ICAO), Safety Management Manual, Document No. 9859, 2017.
4. Hessburg, J. 2001. *Air Carrier MRO Handbook*. McGraw-Hill. New York. ISBN 0-07-136133-2.
5. FAA, 2017, accessed June 21, 2017, Website: https://www.faa gov/regulations_policies/handbooks_manuals/aircraft/amt_handbook/media/FAA-8083-30_Ch12.pdf.
6. FAA, 2017, accessed June 24, 2017, Website: https://www.faa.gov/aircraft/air_cert/continued_operation/ad/.
7. Moubray, J., 1997. *Reliability Centered Maintenance 2*. Industrial Press. New York. ISBN 0-8311-3146-2.
8. Nowlem, S. and Heap, H., Reliability Centered Maintenance Report Number AD-A066579, United States Department of Defense, 1978.
9. Piotrowski, D., "Validation of a Structural Health Monitoring (SHM) System and Integration into an Airline Maintenance Program," *ATA NDT Forum*, San Francisco, CA, September 23, 2014.
10. Aviation Week Network, "2017 MRO Market Share by Category," July 12, 2017.
11. Cameron, D., "Boeing Eyes Higher Sales in Parts and Services," *Wall Street Journal*, article published June 19, 2017.
12. Wilmering, T.J. and Ramesh, A.V., 2005, "Assessing the Impact of Health Management Approaches on System Total Cost of Ownership." *IEEE Aerospace Conference*, 1-4: 3910-3920.
13. Benedettini, O., Baines, T.S., Lightfoot, H.W., and Greenough, R.M., "State-of-the-Art in Integrated Vehicle Health Management," *Proceedings of the Institute of Mechanical Engineers, Part G: Journal of Aerospace Engineering*, 223 2009: 157-170.
14. Rolls-Royce, "Rolls-Royce Celebrates 50th Anniversary of Power-by-the-HourTM," Rolls-Royce Press Release, dated October 30, 2012.

15. RollsRoyce, 2017, accessed June 7, 2017, Website: https://www.rolls-royce.com/products-and-services/civil-aerospace/aftermarket-services.aspx.
16. GE Aviation, 2017, accessed June 7, 2017, Website: https://www.geaviation.com/commercial/truechoice-commercial-services.
17. Boeing, 2017, accessed June 8, 2017, Website: http://www.boeing.com/services/index.page#/commercial-services.
18. Pratt & Whitney, 2017, accessed June 8, 2017, Website: http://www.utc.com/News/PW/Pages/Pratt-Whitney-Launches-EngineWise-Service-Brand.aspx.
19. Shay, L.A., "Airbus Expects More MRO Market Consolidation Expands MRO Alliance," MRO-Network, article dated October 4, 2017a.
20. Boeing, 2006, accessed June 8, 2017, Website: http://boeing.mediaroom.com/2006-05-01-Boeing-to-Acquire-Aviall-to-Enhance-Its-Growing-Services-Businesses.
21. Shay, L.A., "Delta's Predictive Analytics Drives On-Time Performance Improvements," MRO-Network, article dated October 27, 2017b.
22. Delta, 2015, accessed June 8, 2017, Website: http://www.deltatechops.com/mro-capabilities/view/category/rolls-royce-approved-maintenance-center.
23. Delta, 2017, accessed June 8, 2017, Website: http://www.deltatechops.com/PDF/dto_info.pdf.
24. Lufthansa Technik, 2017, accessed June 8, 2017, Website: https://www.lufthansa-technik.com/.
25. Aviation Partners Boeing, 2017, accessed June 8, 2017, Website: http://www.aviationpartnersboeing.com/about_history.php.
26. Airbus, 2017, accessed June 8, 2017, Website: http://www.airbus.com/newsroom/press-releases/en/2016/05/delta-air-lines-becomes-launch-customer-of-airbus-digital-prognostics-and-risk-management-solution.html.
27. Airbus, "Health Monitoring and Prognostics," 2016, Airbus FAST Publication #58, accessed June 8, 2017.
28. U.S. Patent Office, U.S. Non-Provisional Patent Application No. 15-398,812, 2017.
29. Hokura, E., "Compliance RFID Maintenance Program at Delta," *3rd RFID and Paperless Aircraft Operation Conference*, Seattle, WA, November 15, 2016.
30. Andresen, J., *Aviation RFID in Perspective* ATA e-Business Forum, Amsterdam, June 13, 2017.
31. Lambda Technologies, 2017, accessed June 8, 2017, Website: http://www.lambdatechs.com/pdf/STCRelease.pdf.

CHAPTER 9

Safety: Culturally Embedded

Robert Voros
Textron Aviation Inc.

> *You've got to expect things are going to go wrong. And we always need to prepare ourselves for handling the unexpected.*
>
> *Neil Armstrong, 2005,* Magnificent Desolation: Walking on the Moon

Safety

In civil aviation, safety is a culture. From initial aircraft concepts to the words of a flight attendant, safety is a key consideration in our daily decisions. This influence is driven by individuals, governments, companies, and organizations reacting to a dramatic history of accidents and technical innovation. The results are found throughout the other chapters of this book. Regulations, standards, testing, and maintenance all come back to one idea—keeping civil aviation safe.

This chapter doesn't seek to summarize what you will find in the other chapters. Because safety is a motivation throughout the industry, one can delve any topic to find the safety inspirations which brought about each. It is difficult to point to any one regulation, standard, or practice that defines safety, because it is a combination of activities happening simultaneously which produces the results we currently experience. Safety is part of the design, build, and operation of each aircraft. It is the focus at every airport and air traffic control center. In 2017, this culture brought us to have only 1 fatal flight in every 12 million flights of large airplanes in commercial transport service [1]. In the context of the nearly 35 million large, commercial flights which happened around the globe that year, this statistic is a stunning result of a combination of organizations, processes, and people working together with a common consciousness. The industry adapts and responds to events which show us how and where safety is compromised, so that we can correct it. Aviation will never be without risk or a chance of fatal incident, but the industry works continuously to reduce the probability of that occurrence.

Within this environment, there is a unique thread which exemplifies a way of thinking about the topic. Safety assessment is a method, which is expressed in regulations and standards, implemented by aircraft developers and manufactures, and used to sustain the safety of the aircraft for its entire life. It works by providing methods which reduce the probability of fatal accidents by driving the reliability of aircraft designs. The concepts which make up this practice have been maturing since the early days of civil aviation and have grown with technology to become safety practices used for every type of air vehicle. The story of this thread is one that reflects the evolution of safety concepts in the industry and one that can tell us about its future.

The Origins of Safety Assessments

For decades, aviation technologies, regulations, and analysis have coevolved to bring us the level of safety we expect today. Over the past half century, we have seen growth not only in the technology which has brought us the modern aircraft but also in the way we have come to think about how that technology does that job safely. As the aircraft advanced from the rudimentary mechanical systems of its earliest days to the powered automation of today, our ability to comprehend the new risks and complexities of these systems also advanced.

In the beginning of civil aviation, the complexity of systems was limited by the technologies of the day. The civilian vehicles of the 1940s and 1950s were often reciprocating engine airplanes with manually operated mechanical systems in which the human was the primary decision-maker. Despite their simplicity, these aircraft could crash due to single failures within their systems, particularly engines. As the demand for air travel grew, the motivation to address these design issues likewise developed. In 1944, the Convention on International Civil Aviation was held in Chicago and began the formalization of agreements to provide equivalent standards between nations regarding the design and operation of aircraft. By March of 1949, the Standards and Recommended Practices for the Airworthiness of Aircraft was adopted and designated Annex 8 by the International Civil Aviation Organization (ICAO). These requirements established broad standards for aircraft design intended to facilitate nations developing their own standards and regulations for aircraft safety [2].

While accidents were reduced following this regulatory advancement, subsequent accidents revealed the contributions from multiple failures. Early in the 1950s, the "fail-safe" concept developed from the principle of preventing single failures. A fail-safe design allows a system to continue to provide functionality after a failure somewhere within that system or to give the flight crew time to address the issue. Specific fail-safe criteria were directed through regulations to the design of certain systems (e.g., engines, brakes, and flight controls). These specific regulations on critical systems drove aircraft design to continue to provide functionality after a single failure or other defined number of failures. In addition to these system-specific criteria, the introduction of Civil Air Regulation (CAR) § 4b.606 identified a more global approach. CAR §4b.606 described a more general, aircraft-level context regarding the reliability of functions on the aircraft. It caused designers to postulate possible hazards resulting from the failure of an aircraft's functions or from malfunctioning [3]. This more general type of regulation represents a step away from the implementation-specific language of existing regulations and a step toward the broader consideration of the total aircraft, opening the design space for the aircraft's systems and overall safety.

As electricity found its way into more aircraft systems later in the 1950s, the regulations took on greater considerations of the relationship between systems, particularly with respect to their failures [3]. Tools such as Failure Modes, Effects, and Criticality Analysis (FMECA) developed in military procedures, MIL-P-1629 [4], were being applied in civil aviation to identify and understand the effects of these failures. At this time, these fail-safe requirements relied heavily on "engineering judgment" and qualitative methods [5]. As aircraft added more systems and functionality through the 1960s, they also became more critical, integrated, and complex. It became increasingly difficult to simply duplicate functions through redundant systems and qualitatively determine that the potential hazards on those aircraft were addressed.

These redundancies also lead to system failures being masked or latent. Latent failures are those which are not obvious in normal operation. These failures emerge in situations where the redundancy meant to protect the aircraft, by continuing to provide a function after a failure, masks failures in the system which would take away that protection. Unknown to the crew, these latencies leave the aircraft one failure away from catastrophe [5]. As aircraft systems increasingly utilized technologies such as hydraulic and electrical power, flight crews became less directly aware of the status of the viability of their aircraft's systems. Unlike a cable-operated flight control system, which physically connected the pilot to the control surface on the wing or tail through steel cables, a fully powered flight-control system meant that failures could occur between the pilot and control surface and the pilot would not be aware until the aircraft failed to respond to a critical command. The risk of these latent critical failures began driving aircraft designers to identify them and provide new warnings in the cockpit. Complex systems, latent failures, and warnings to pilots were some of the impetuses that motivated the next advancement in design safety.

The *Tentative Airworthiness Standards for Supersonic Transports* of 1965 further developed concepts to analyze systematically the growing complexity driven by this application's increase in functionality, redundancy, and warnings [6]. The Federal Aviation Administration (FAA) updated its regulations to drive toward a more comprehensive, systematic analytical approach [7]. This update to Federal Aviation Regulation § 25.1309 (formerly CAR § 4b.606) ushered in many of the concepts which formed the modern methods of system safety assessment to be used for showing compliance to the rule. These methods of safety analysis worked to provide low incident/accident rates while acknowledging that the layers of redundancy introduce costs, complexity, and the risk of new failure conditions which could be unseen. The introduction of a quantitative approach to safety analysis was prescribed to show that sufficient coverage was provided by designers while limiting the risks inherent in unnecessary complexity. At the same time, concepts such as the FMECA were being further developed by the National Aeronautics and Space Administration (NASA) for the Apollo program [8] and applied in civil aviation to identify and understand the effects of these failures. By 1967, the Society of Automotive Engineers (SAE International) had published SAE ARP926 which broadened the concepts of the military FMECA concept to consider these more integrated and complex concepts [9].

The Introduction of Design Assurance

By the late 1970s, aircraft designers were looking to solve the challenge of providing redundancy by using volume saving digital systems. As software found its way into critical systems, the industry and regulators began developing guidance on how regulatory requirements for such augmented systems might be met. At this time, software was used in conjunction with analog systems to protect against faults in the software. In May 1980, the Radio Technical Commission for Aeronautics (RTCA) Special Committee 145 for "Digital Avionics Software" was tasked by regulators to (1) develop a plan and procedures by which the validity of software may be demonstrated, (2) recommend minimum standards for consideration in digital avionics to implementation, and (3) maintain coordination with the European organizations with aligned interests. The industry and regulators saw a future where software would be directly responsible for all paths of critical functionality on an aircraft, and guidance was necessary to ensure software would exhibit equivalent safety to their analog predecessors.

The nature of digital systems is not directly comparable to their analog counterparts and, therefore, requires a different approach. Special Committee 145 published RTCA DO-178, *Software Considerations in Airborne Systems and Equipment Certification* on November 18, 1981. It addressed the challenge of providing the equivalency to analog systems by describing a disciplined approach to software development through recommending methods for design, testing, configuration management, and documentation of that software [10]. As part of that guidance, DO-178 developed the concept of categorizing functions based on the effects of design errors. This categorization was comparable to MIL-STD-1629A [11], an update to MIL-P-629, which categorized the effects of failures. While failures are typically considered the loss or deterioration of components, errors are a result of the development process. While a system may malfunction because a component has broken, a pristine system may malfunction because it was not designed as intended due to errors in its initial requirements or in failing to confirm that what was implemented in the physical vehicle matches those requirements. This concept was matured in subsequent revisions of DO-178, and eventually aligned with core concepts in the recommended practices for the safety assessment of the aircraft.

By the 1990s, the regulations and guidelines for small and transport category airplanes, rotorcraft, and engines began reflecting this merging of development assurance and safety assessment concepts. The resulting process includes the systematic methods by which an aircraft's functionality is evaluated to determine potential conditions which might affect the aircraft and its occupants, which results from one or more failures or errors. These failure conditions drive the development of safety objectives which are used to establish requirements whose implementation can be verified in the final design. The safety assessment process is a consideration from the beginning of an aircraft's design, and is updated and iterated as that design matures over its operational life. In the end, the safety assessment process demonstrates that the objectives set for the aircraft at the beginning or through updates were met in the final design. This assessment is qualitative in many respects but relies on quantitative analysis to support these qualitative statements to show the objectives were achieved. The process relies on systematically reviewing the aircraft, its functionality, and its operating conditions to qualitatively identify and classify what failure conditions the vehicle might face. While the nature of these conditions and effects cannot be quantified, their safety criticality can be given a classification of severity. The relationship between the severity of the failure condition and the probability of its occurrence should be inverse—the more severe, the less probable it should be. This probability, then, can be quantitatively evaluated relative to the failure rates of components in tools such as Fault Tree Analysis and Failure Modes and Effects Analysis (FMEA). These quantitative analyses are used to support the qualitative assessment that the failure condition has been mitigated. The safety assessment process provides a methodology to assure that each aircraft is assessed comprehensively for safety from the beginning, and meets the requirements that assessment determines when it is put into service.

The process that follows is applied to the development of most aircraft types, from the large, commercial transport to small airplanes and helicopters. While much of the discussion involves transport category aircraft, the process itself is flexible. By defining a method that looks at the vehicle and systems based on their functionality, one will find the processes are tailorable to a variety of applications. The consistency they present is a uniform focus on limiting the probability of safety hazards in every design toward a continuum of aviation safety.

The Safety Assessment Process

As civil aviation was maturing, it continued to acquire concepts from military standards and practices. This borrowing was routine for companies producing both military and civilian products as they sought to keep organizational processes common. In 1994 this sharing significantly broadened. The Department of Defense shifted its approach to standards in the June 1994 "Perry Memorandum" [12], which directed the department toward "moving to greater use of performance and commercial specifications and standards." This memorandum brought about the 1995 cancellation of the 1974 military standard, MIL-STD-499A, *Engineering Management*, which described the planning for system engineering processes, and grew the library and influence of civilian systems engineering standards.

By 1996, the influence of systems engineering is reflected in the concepts of SAE International Aerospace Recommended Practice (ARP) 4754, *Certification Considerations for Highly-Integrated or Complex Aircraft Systems* [13], and ARP 4761, *Guidelines and*

Methods for Conducting the Safety Assessment Process on Civil Airborne Systems and Equipment [14]. While ARP 4761 primarily provides a compilation of the practices for a safety assessment process for transport category aircraft, the processes it describes are also applied to other aircraft types. It embodies complex system analysis and functional analysis which support the development of software and hardware. It describes a process to systematically analyze the entire aircraft for potential failure conditions, categorize those conditions by the severity of their effects, identify which conditions require quantitative analysis, and verify the aircraft was properly assessed.

Commonly called upon as guidance for showing compliance to their regulations by the FAA, the European Aviation Safety Agency (EASA), and other certification authorities, the safety assessment process described by ARP 4761 provides guidelines for assessing an aircraft in a systematic way to reduce an organization's reliance on engineering judgment. The A revision of SAE ARP 4754, *Guidelines for Development of Civil Aircraft and Systems* [15], describes how that process may be implemented consistently and comprehensively throughout an entire program. The safety assessment process begins at the aircraft-level assessment which considers the operating environment of the aircraft and the functionality of its systems. This process works to address the contribution of failure of interrelated systems to these failure conditions and confirm that the risks which these failures impose are mitigated.

In alignment with ARP 4754A and systems engineering practices, the safety assessment process is conducted in two major phases—the top-down Safety Requirements Development and Validation phase and the bottom-up Safety Requirements Verification phase (see **Figure 1**). The top-down phase begins with developing safety requirements starting at an aircraft level. The process begins early in an aircraft's development and starts with consideration of how the functions, from the perspective of the aircraft as a whole, might fail or malfunction. By reviewing how systems might be used to mitigate these failure conditions, the process allocates high-level safety requirements from the

FIGURE 1 High-level safety assessment process.

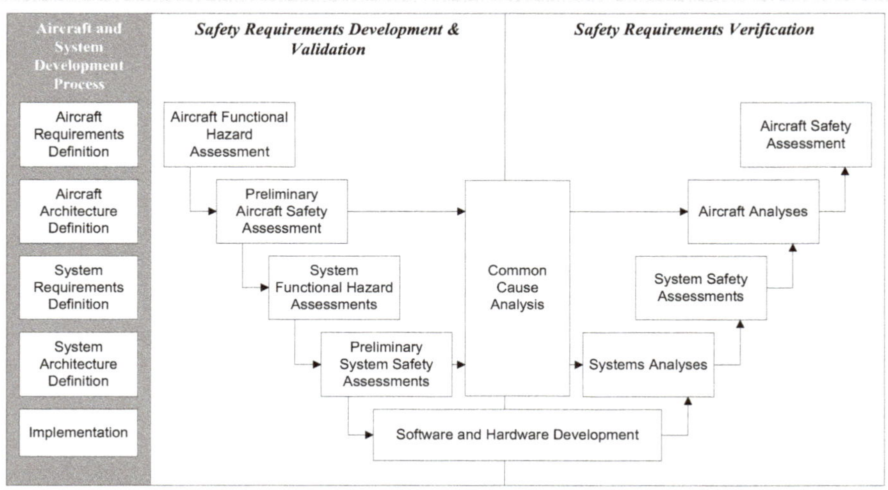

assessment to systems, and in turn are similarly assessed, after making further design decisions, down to the software and hardware. The bottom-up phase analyzes the detailed design of the aircraft starting with the results of the software and hardware implementation against the safety requirements from the earlier phase. The bottom-up analyses are initially conducted at the system level to demonstrate that the safety requirements were met. It considers not only how the functions are affected by failures and malfunctions but how threats to the aircraft from external sources, such as a bird strike, or physical threats from the aircraft's own systems and equipment, such as explosions, fire, and heat, may also compromise the vehicle's safety. The process concludes when the aircraft is looked at as a whole to confirm that the top-level safety requirements were likewise satisfied.

For brevity, the following sections describe in a linear fashion the high-level steps of a safety assessment process. It should be recognized that the process is iterative throughout an aircraft's operational life, from initial development through its field experience, and requires updates as more knowledge is gained about the design. Because the process starts at the aircraft level and subsequently moves to the system, software, and hardware levels, safety issues discovered during detailed design activity, later in the program, may have implications on the higher aircraft and system levels of design. As part of the Development Assurance Processes described in ARP 4754A, the system, software, and hardware developments are often checked against their next higher level to assure that no objectives or requirements from those higher levels were violated by changes in those lower levels. Processes such as configuration management are used to provide a mechanism to communicate and control changes during development to assure that the data used are the latest and the most appropriate. The documentation of these processes can be varied so long as the intent of the process is met. ARP 4754A recommends the use of development planning to explain the expectations for the execution of these processes.

Safety Requirements Development and Validation

This top-down phase consists of three major types of safety assessments conducted at the aircraft and system level. Functional Hazard Assessments are processes which look at the vehicle's functions at the aircraft and systems levels to determine conditions in which failures of those functions might occur, classify those failure conditions, and establish safety objectives from them. Preliminary Safety Assessments performed at the aircraft and systems levels take these safety objectives and use them to assess proposed aircraft and system architectures to determine if they satisfy the objectives. If so, these preliminary assessments then set requirements on the aircraft architecture, systems, software, or hardware. Finally, the Common Cause Analyses build on the Preliminary Safety Assessments to establish safety requirements considering causes not directly related to functional failure analysis, such as physical external threats or physical installation issues that could compromise the redundancy or independence in a system. Together, these tools develop a set of requirements which drive safety considerations in a consistent way from the earliest stages of development and highest level of operations to the lowest level of equipment, software, and hardware.

Aircraft Functional Hazard Assessment: The safety assessment process begins with an Aircraft Functional Hazard Assessment, which is a systematic evaluation of the aircraft-level functions. Because this process begins early in development, an aircraft-level function does not need to recognize the detail of systems used to implement those functions. An aircraft-level function can be as high level as "provide on-ground deceleration" with an accompanying set of requirements which describes the landing distance requirements for the aircraft as a whole. Once defined, the vehicle's functions are reviewed to establish failure conditions that have direct or consequential effects on the aircraft and/or its occupants. These conditions are caused by one or more failures or errors and consider flight phase, operational and environmental conditions, and external events which can make the condition more adverse. Failure conditions individually consider the loss of each function and the unintended operation of each function. These conditions do not directly consider the failure which might bring the condition about, though they do include consideration of the workload of the flight crew to compensate for the event. For each condition, the effects are determined relative to their impact on the aircraft, passengers, and flight crew. The result of the evaluation is a classification of the severity of that effect.

The categories of these conditions have specific terms spanning a range of severities from "no safety effect" to "catastrophic," as described in both regulatory guidance and industry guidelines. The most severe failure condition is classified as catastrophic, which are those that prevent continued safe flight and landing, normally resulting in hull loss and death. In alignment with the concepts in DO-178, these categories are used to set the safety objectives to mitigate those failure conditions. The safety objectives, set in regulatory guidance, establish an inverse relationship between the severity of the failure condition and the probability of its occurrence; see **Figure 2**. The ultimate objective is to assure that an inverse relationship exists between the probability of a failure condition and the severity of its effect on the aircraft and its occupants [2]. In addition to probabilistic values that guidance may associate to these classifications, the most severe classification, catastrophic, often requires that there must be no single failure which can cause such a deadly condition to occur. This "no single failure" requirement enforces redundancy and independence requirements.

The Aircraft Functional Hazard Assessment also describes a plan of how the organization will confirm the failure conditions which it describes were determined to be correct and complete, and how they will be shown through subsequent analysis to be mitigated in the final design.

Preliminary Aircraft Safety Assessment: Once safety objectives are determined at the aircraft level, the proposed architecture of the aircraft is assessed against those objectives. This Preliminary Aircraft Safety Assessment is used to conclude that the preliminary architecture can meet these safety objectives, and to set the safety requirements for the systems involved. This assessment includes a high-level consideration of how system failures might contribute to the aircraft-level failure conditions.

The aircraft architecture describes how an aircraft's behaviors are being provided by its systems' functions and how those systems relate to each other. By assessing this architecture against the safety objectives, the systems developers can begin to determine

FIGURE 2 Probability of failure conditions vs the severity of their effects.

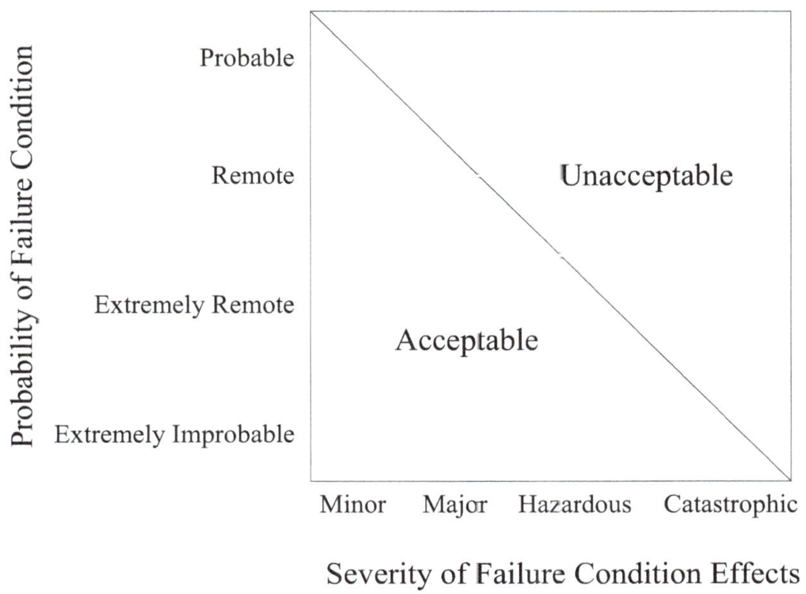

how reliably their systems need to provide that functionality, as well as what kind of critical independencies and redundancies these systems need from each other or need internally to be sure they prevent or do not contribute to critical failure conditions. Crucial to understanding the relationship between systems, this safety assessment provides a top-down view of the relationship of systems to the aircraft failure conditions.

The criteria established by the Preliminary Aircraft Safety Assessment help system developers make fundamental decisions in the development of their systems' functions, architectures, and independence needed from other systems. These criteria are also used to establish guidelines for the development of systems to reduce the risks from common causes which might violate the required independence. The Common Cause Analysis looks at the independence requirements from the Preliminary Aircraft Safety Assessment and to establish design guidelines, such as clearance volumes around components, for circumstances may violate those independences including maintenance errors, physical threats from other systems (heat, leaks, interference, etc.), and other threats. At this high-level stage, a Common Cause Analysis can be used to set requirements which may include installation limitations, zones of the aircraft with special safety considerations, and maintenance practices.

System Functional Hazard Assessment: The Functional Hazard Assessment process is repeated at the system level. In the development process, a system must meet the safety requirements allocated to it by the aircraft architecture to establish a more specific set of functions within its boundaries. For each system, functions are assessed by their System Functional Hazard Assessment process to identify failure conditions specific to that systems functions. The system-level failure conditions are developed systematically and consider the loss of each function and its malfunctions. They do not directly consider the failure which might bring the condition about, though they do consider the flight crew's workload. For each condition, the effects are determined relative to their impact on the aircraft, passengers, and flight crew. The effects determined from these evaluations are also classified based on severity leading to the establishment of safety objectives. While all failure conditions require a qualitative assessment to verify that they have been addressed by the system, critical conditions also require quantitative analysis to more specifically demonstrate that those critical conditions have been satisfied. Regulatory guidance such as AMC 25.1309, AC 25.1309-1A, AC23.1309-1E, and more sets a probability value for these critical classifications which the analysis must meet. Like the aircraft level, failure conditions identified as "catastrophic" require that "no single failure" can cause them. As with the aircraft level, each System Functional Hazard Assessment also describes a plan of how the organization will confirm the failure conditions which it describes were determined to be correct and complete and how they will be shown through subsequent analysis to be mitigated in the final design.

Preliminary System Safety Assessment: Similar to the assessment process at the aircraft level, once safety objectives are determined in the System Functional Hazard Assessment, proposed system architectures are assessed against those objectives in a Preliminary System Safety Assessment. This assessment is used to determine if the preliminary architecture can meet the safety objectives. This preliminary assessment also establishes safety requirements for the components and software. This process is conducted for each system.

The system architecture describes how a system's functions are being provided by its components and software, and how those elements relate to each other. By assessing this architecture against the safety objectives, component and software functionality and independence requirements can be established as needed to prevent or to not contribute to critical failure conditions. These criteria help software developers determine a level of development rigor based on the classification of failure conditions. This level of rigor is used to ensure more critical functionality receives greater scrutiny through development to confirm that the right requirements were used and properly implemented in the final design. To do this, these levels modulate development processes such as those defined in RTCA DO-178. At this level of detail, the criteria also led to the development of aircraft flight manual procedures for the abnormal and emergency procedures that may be used in the event of the failure conditions. The critical conditions at this level require the use of quantitative analysis, which can identify latent failures within systems, and the frequency at which they should be checked through maintenance or preflight procedure. There are many ways to address latent failures including on-aircraft systems tests, preflight checks, and maintenance inspections. The quantitative analyses help allocate failure probabilities to components and set the intervals for maintenance inspections, which support finding these latent failures. The assessment of the system architecture also supports the development of the levels of rigor by which software and airborne electronic hardware are developed. In 2000, a process for developing airborne electronic hardware similar to DO-178 was provided in the RTCA document DO-254, *Design Assurance Guidance for Airborne Electronic Hardware* [16].

The system-level safety requirements help system developers make fundamental decisions in the development of their components, software functions, detailed architecture implementations, and independence needed within the system. Like the aircraft level, these criteria are also used in the context of Common Cause Analysis to establish specific requirements for each system to reduce the risks of causing failure conditions based on installation and threats from outside of the system such as maintenance errors, physical external threats from other systems (heat, leaks, rubbing, etc.), and other threats which might violate the necessary independence within the system. The equipment, software, and airborne electronic hardware design processes are conducted according to the level of rigor determined through the safety assessment process.

Software and Hardware Development Processes

A critical aspect to the development of the modern safety assessment process was its relationship to the software development process. Preliminary System Safety Assessments provide numerous safety requirements to software and hardware development processes. Among these requirements are Item Development Assurance Levels (IDALs) which are used to establish the level of rigor used in the development of that software and hardware. The level of rigor which the IDALs help set describes the types and amount of development activities which should be conducted on that software or hardware to minimize the occurrence of errors, which may result in losses or malfunctions. The activity which uses these IDALs is defined in standards such as DO-178 and DO-254. Just as the probability of occurrence of a failure condition should decrease with the severity of that failure condition, the risk of an error in the development of software and hardware should be likewise decreased. These IDALs associate to the failure conditions to assure more development activity is performed for functions developed to mitigate more severe failure conditions. This increased activity reduces the risk of an error occurring in the design of the function through that software or hardware process.

For complex and highly integrated systems, this development assurance level concept is also applied, as described in SAE ARP 4754A. This process allocates Functional Development Assurance Levels (FDALs) to the development process of the functions within these systems. Similar to the software and hardware processes, the ARP 4754A processes describe how the FDALs are used to establish rigor in the development of those system functions to assure that errors to processes feeding the software and hardware design processes are minimized with quantitative or qualitative criteria.

Safety Requirements Verification

As the design matures and the software and hardware development progresses, implementation details are analyzed against the requirements developed in the safety assessment process. This bottom-up process begins implementation of the verification phase which focuses on demonstrating that the requirements set by the processes in the requirements development and validation phase have been met by the implemented design.

The processes used to verify the implementation of these requirements utilize the preliminary assessments from the preceding phase as a foundation. Like the validation phase, the verification phase can be iterative and drive requirements or design changes.

System-Level Analyses: The safety analysis for the verification of systems begins at the detailed level. Software and hardware are matured in their development processes and are ready to be integrated in the system installation. These items are analyzed to assure they met their requirements and they were developed to the level of IDAL rigor required by the Preliminary System Safety Assessment(s). FMEA and FMECAs are conducted on components to determine how they might fail and the consequences of those failures on the next higher assembly. For critical conditions which require quantitative analysis, the effects of these component failures and their probabilities are analyzed against the failure conditions through tools like fault tree analysis to see how these failure modes might contribute to the failure conditions determined early on in the program's development. Both FMEAs and fault tree analysis are described in SAE ARP 4761 as well as civilian and military standards including SAE ARP5580 and MIL-STD-1629A.

Common Cause Analyses: Common Cause Analyses establish physical and functional separation, isolation, and independence requirements between systems and items and verify that these requirements have been met. Principles of these analyses were used in the Preliminary Safety Assessments, mentioned above, in the requirements development and validation phase to set specific requirements on systems. On the verification side of development, these processes are used to verify that the requirements from those assessments have been met. Common Cause Analysis is a generic term encompassing the processes of Zonal Safety Analysis, Particular Risk Analysis, and Common Mode Analysis.

A Zonal Safety Analysis analyzes specific volumes of the aircraft to look for physical violations of independence due to the failures, interferences, and installation challenges posed by other systems. Such violations would not be apparent in a functional safety analysis like a fault tree analysis nor a failure modes and effects analysis. The Zonal Safety Analysis is a spatial analysis that ensures that the equipment installation meets the safety requirements with respect to its basic installation, interference between systems, and maintenance access needs. It helps identify if there is sufficient clearance around parts so that in different operating conditions, parts won't damage the structure or other parts in their surroundings. Interference and damage can occur to surroundings in normal operation due to component vibration, heat production, or sag under normal load. It also considers the physical effects of equipment failures and maintenance errors with respect to neighboring systems and structures. Spatial effects of system failures include the effects of a leak of a hydraulic line or the failure of a high-pressure duct failure on the surrounding systems and structure. This analysis is conducted throughout the development of the aircraft, starting with computer models, progressing to test articles, and finally being conducted on a production vehicle.

Like the Zonal Safety Analysis, a Particular Risk Analysis looks for violation of physical independence due to external threats that are distinct from system failures, errors, and maintenance. A Particular Risk Analysis analyzes events or influences which are external to the aircraft or within the aircraft but external to the system(s) and item(s) being analyzed but which may violate safety requirements. These risks often are defined by other regulations and regulatory guidance including topics such as bird strike (14 CFR 25.631 and 14 CFR 25.571), tire burst and flailing tread (14 CFR 25.729(f)), and High-Intensity Radiated Fields (14 CFR 25.1317).

A Common Mode Analysis is performed to verify that identified independence requirements are met in the final design. This analysis evaluates the implemented design for circumstances that defeat redundant design principles: implementation, manufacturing, and maintenance errors and failures of system components. Not only does this analysis help confirm necessary independence, it may identify new failure conditions due to the discovery of previously unconsidered common failures. Common modes include obvious, shared elements such as the use of a common electrical or hydraulic power supply to less direct considerations, such as the independence necessary for software development.

System Safety Assessment: Each System Safety Assessment collects, analyzes, and documents the verification data that show the implemented system meets the safety requirements established in the Preliminary System Safety Assessments. This bottom-up approach evaluates each failure condition identified in the System Functional Hazard Assessment according to the analysis methods it specified (qualitative and quantitative analysis as necessary). This assessment shows how failures combine to lead to a specific failure condition, and that those failure conditions are mitigated by the implemented design. Probabilities are used for quantitative analysis to show that the systems functions have sufficient reliability to meet the safety objective. Quantitative analysis also shows that there is sufficient redundancy to continue to provide critical functionality after certain failures. The assessment evaluates the Common Cause Analysis to confirm that the implemented design did not violate any of the independence requirements. Maintenance documentation is examined to show that any mitigations by maintenance tasks are addressed, and that intervals for those maintenance tasks are supported by the quantitative analysis which drove them. Aircraft flight manuals are likewise reviewed to confirm that the procedures deemed necessary to mitigate certain failure conditions are properly documented.

Aircraft Safety Assessment: The Aircraft Safety Assessment collects, analyzes, and documents the verification data that show the implemented systems meet the safety requirements established in the Preliminary Aircraft Safety Assessment. This bottom-up approach evaluates each failure condition identified in the Aircraft Functional Hazard Assessment. This assessment shows how failures combine to lead to a specific failure condition and that those failure conditions are mitigated by the implemented designs in the systems. The assessment evaluates the Common Cause Analysis to confirm that the implemented design did not violate any of the independence requirements set at the aircraft level.

Safety Management and the Safety Life Cycle

The safety objectives for the aircraft apply to its entire life cycle. The previously discussed safety assessments not only provide confidence that there is sufficient redundancy, availability, and integrity for the aircraft's functions but that the maintenance tasks and flight manual procedures support and reflect safe practices. As the aircraft gains more service experience, operators may discover new failure conditions or modes of operation not previously considered. These discoveries would cause a reconsideration of the aircraft's safety assessment to identify if a new failure condition or design issue has emerged. This feedback to the safety assessment would reveal if a field action may be necessary.

In the 2000s, the ICAO continued to mature the systematic safety concepts for the design and certification aspects of the vehicle to include its entire life cycle. The concept of the Safety Management System (SMS) was developed to proactively address safety risks that may emerge from the predicted growth in air traffic over the next 15 years. The risks approached at the SMS level focus on assuring a continuity of safety assurance for the products in the aviation environment and less directly about an aircraft's design-level safety risks. It does this by assuring those organizations that are responsible for the type definition, production, and operation of aircraft provide an organization-wide approach to manage the product's safety risk in a formal, top-down way for its entire life. The SMS guides organizations to describe and understand the extent of their processes which can affect the product's safety [17]. This concept assures that the processes and procedures are in place so that there is confidence that vehicles are meeting their safety needs throughout their life cycles. These companies already contain many of the safety management provisions, but SMS bring them together in a comprehensive operating environment.

The SMS views the company and its processes much the same way as the safety assessment process views systems on an aircraft. Like the aircraft and system-level perspectives in the safety assessment process, the SMS looks at the top-level operating environment of the organization and the processes (systems) inside of it which provide the safety functionality. For a holder of type certificates and production certificates, these systems are categorized into three main areas: compliance with airworthiness regulations ("compliance assurance"), compliance with quality regulations ("conformity assurance"), and continued operational safety (COS) ("safety assurance"). These three concepts are more simply thought of as the specific phases of the aircraft's life cycle: design and certification, production, and continued airworthiness (**Figure 3**).

Design and Certification

These processes are all those used to design and certify a product that is safe and compliant to the regulations. This process includes the previously described safety assessment process as well as the more aircraft- and system-specific requirements and regulations which also provide safety objectives, such as those for flight controls, engines, fuel, and so on. "Compliance assurance" refers to the ability to show that the final design

FIGURE 3 Safety management system operating environment.

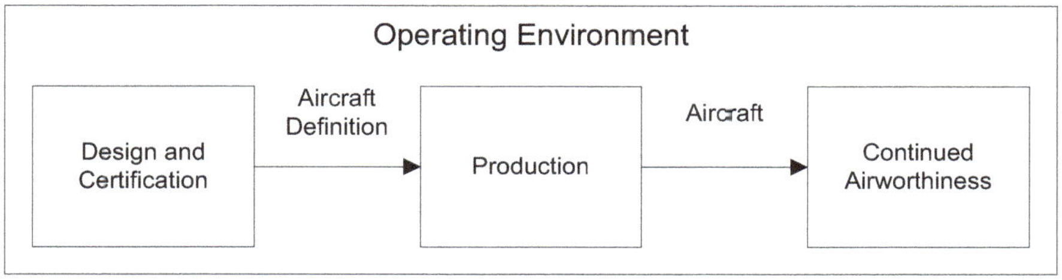

meets the safety and regulatory requirements placed on it. Ultimately, this is shown by successful receiving a type certificate for the aircraft's design. Issues which emerge in production or in fielded aircraft may also feed back to the design and certification process for design changes.

Production

Production processes that help assure safety of the aircraft are those that show that the "as built" aircraft matches with the definition which is type certificated. Defects, errors, and other deviations from the production design may have implications regarding the safety of the product. The conformity assurance process monitors for any deviations to the aircraft's design and the production processes during the aircraft assembly. Deviations, once found, are evaluated for potential safety risks against the product design and certification. The impact of these deviations is assessed against the product line and the fielded aircraft, which may have been affected, to determine the corrective actions which may need to be taken.

Continued Airworthiness

The COS is sustained by the monitoring of fielded aircraft to determine if safety-related issues have emerged. The root cause of fielded aircraft safety issues can come from a variety of sources including maintenance, manufacturing, the environment, and design. Fielded aircraft are monitored to determine if an accident incident, or even customer compliance implies a larger issue with the fleet. The COS can feed back to the production process as well as the design and certification process for corrective actions.

Operating Environment

The combination of these processes into the operating environment of an organization is driven by a core set of organizational safety objectives for the organization's control of their vehicles' life cycles. This common source provides a link between all processes to assure the visibility to aircraft issues and responsiveness to resolve them. Since these processes already have many interactions, an SMS provides uniformity and continuity between the processes. The SMS also drives deeper consideration of organizational and process changes which may have an impact on aircraft safety. Similar to the safety assessment process for aircraft, the organization's processes are assessed for potential safety risks, and if necessary, appropriate controls are implemented.

The Future of Safety

The practice of safety assessment has evolved over the past half century from failure mode analysis, to the current broad, methodical approach of safety assessment, and on to the SMS approach to organizational considerations of safety. The industry continuously works to reduce the probability of these events though we can never guarantee that there will be zero. The methods discussed in this chapter have continued to improve and enhance safety from the earliest days to present, advancing with every technical leap. They are applied to all aircraft types to assure a continuum of safety throughout our industry from airliners to small helicopters. The flexibility of these processes presents an opportunity to align safety expectations with new technologies which have yet to gain full commercial use such as space vehicles, unmanned aircraft systems (UAS), etc.

New technologies do pose new questions for safety assessment and the continuum. For instance, a tablet interface with an autonomous, passenger-carrying flying machine exhibits a new set of considerations for the safety of those passengers. Yet, these considerations can still be discovered and assessed on the fundamental questions we ask about our conventional aircraft, today. What is the effect of the loss or malfunction of this defined behavior (autonomous or not)? What can contribute to bring these failure conditions about? The novelty of the interface does not change the fundamental concerns of the safety risk of its functionality. However, advances continue in the realm of understanding and improving the human involvement in our systems. Human factors regulations, standards, and practices have been more recently evolving toward a more integrated interface with safety assessment. Understanding the role of human behavior in decision-making is involved in processes such as Functional Hazard Assessments and the development of flight manual procedures. Unlike the clearly defined set of behaviors of systems, those of humans are still being studied and understood for their roles in our future world of automatic vehicles of all types. We could see a future where our common preferences and biases are assessed similar to that of our systems toward a safer aviation industry.

As discussed, the trend with these processes has been toward process-driven, systematic methods to consider, address, and verify the safety risks of vehicles from the failure of components to failures within the operation of the organizations which develop and operate them. Safety processes, such as those discussed, also provide an opportunity to simplify the system-specific regulatory environment which persists since the 1940s. In all, the safety assessment process developed in aviation has supported significant advances in aviation technology and is structured to continue.

References

1. Young, A., "to70's Civil Aviation Safety Review 2017," January 1, 2018, Retrieved from to70, http://to70.com/to70s-civil-aviation-safety-review-2017/.
2. International Civil Aviation Organization. (2010). *Annex 8, Airworthiness of Aircraft.*, (11th edn) Vol. 8). Montréal: International Civil Aviation Organization.
3. Federal Aviation Agency, Civil Aeronautics Manual 4b: Airplane Airworthiness, *Transport Category.* (N.E. Halaby, Ed.) (Washington: Federal Aviation Agency, 1962).
4. U.S. Department of Defense. (1949). *MIL-P-1629, Procedures for Performing a Failure Mode, Effects and Criticality Analysis.* Washington: U.S. Department of Defense.
5. Aviation Rulemaking Advisory Committee Systems Design and Analysis Harmonization Working Group. (2002). *Task 2 - System Design and Analysis Harmonization and Technology Update.* Washington: Federal Aviation Administration.
6. Federal Aviation Administration, Tentative Airworthiness Standards for Supersonic Transports, 1965.
7. Federal Aviation Administration. (1964). *Recodification and New Part 25.* Washington: Federal Aviation Administration, Department of Transportation.
8. National Aeronautics and Space Administration. (1966). *RA-006-013-1A, Procedure for Failure Mode, Effects and Criticality Analysis (FMECA).* Washington: National Aeronautics and Space Administration.
9. SAE International. (1967). *ARP 926, Design Analysis Procedure for Failure Mode, Effects and Criticality Analysis (FMECA).* Warrendale: SAE International.
10. Radio Technical Commission for Aeronautics Inc. (1981). *DO-178, Software Considerations in Airborne Systems and Equipment Certification.* Washington: Radio Technical Commission for Aeronautics Inc.
11. U. S. Department of Defense. (1980). *MIL-STD-1629A, Procedures for Performing a Failure Modes, Effects and Criticality Analysis.* Washington: U.S. Department of Defense.
12. Perry, W.J. (1994). *Specifications and Standards - A New Way of Doing Business.* Washington.: Department of Defense.
13. SAE International. (1996). *ARP 4754, Certification Considerations for Highly-Integrated or Complex Aircraft Systems.* Warrendale: SAE International.
14. SAE International. (1996). *ARP 4761, Guidelines and Methods for Conducting the Safety Assessment Process on Civil Airborne Systems and Equipment.* Warrendale, PA: SAE International.
15. SAE International. (2010). *ARP4754A, Guidelines for Development of Civil Aircraft and Systems.* Warrendale: SAE International.
16. Radio Technical Commission for Aeronautics Inc. (2000). *DO-254, Design Assurance Guidance for Airborne Electronic Hardware.* (R. SC-180, Ed.) Washington: Radio Technical Commission for Aeronautics Inc..
17. Aerospace Industries Association of America Inc. (2016). *Safety Management System Practices for Design and Manufacturing.* (USA: Aerospace Industries Association of America Inc.)
18. European Aviation Safety Agency, (n.d.), "AMC 25.1309 Equipment, Systems and Installations EASA Acceptable Means of Compliance."

CHAPTER 10

The Environment: Emissions by Consensus

Christer Hellstrand
Christer Hellstrand Consulting

Introduction

In operation, civil aircraft interact with their environment in mostly undesirable ways. Two of the main negative environmental impacts of aircraft are emissions from their engines and noise from both engines and the airframe. Emissions are a natural by-product of burning hydrocarbon fuel. When a hydrocarbon burns in the presence of air, oxides of carbon and nitrogen are produced. These are unwanted side effects and add to global warming. Having said that, the aerospace industry is unique in its collective agreement to limit these emissions and its technical resolve to achieve this.

There are a number of other sources of emissions. As well as the oxides mentioned above, engines also emit heat and particulates. Also, when an aircraft is viewed on a clear day, distinct white lines can often be seen following it. These lines are termed contrails and are water vapor that is condensed out as it leaves the engine exhaust. Further, when viewed holistically, there are other side effects of flight. The emissions produced in the supply chain by all the manufacturing processes that have to be completed for a plane to come to life could also be examined, although scoping studies have shown these to be relatively very small. These issues, although recognized as important, are not discussed further in this chapter; instead the focus is on emissions and noise in flight.

The chapter is arranged as follows: First, the climate change problem is discussed and the industry response, along four different avenues, examined. Next, each of the ways to reduce greenhouse gas (GHG) emissions is discussed in some depth. The following section details the noise problem and what is being done to address it.

Climate Change

In the Introduction to this book, passenger numbers and the economic effect of the civil aerospace sector on the economy were outlined. Growth, forecast to be 5% per year for the next 20 years, is assured provided the industry can continue to solve the problems facing it and resolve new challenges. One of the biggest recent challenges it faces is that of climate change, linked to GHG emission at high altitude. The best way to approach this problem is from within the industry, as it is the aerospace industry that has the greatest insight and the strongest motivation to address this to continue to grow and prosper.

According to the UN's Intergovernmental Panel on Climate Change report from 2007 [1], aviation accounts for approximately 2% of total (natural and anthropogenic) GHG emissions, a figure that has remained relatively constant since then. However, this estimate could grow higher as nations reduce GHG emissions to prevent catastrophic and irreversible climate change, most countries being in accord with the 2016 Paris Climate Agreement and committing to a significant reduction in GHG emissions. Forecasts are that, by 2050, aviation could account for up to 5% of total GHG emissions.

It is likely that the aviation industry will not continue to grow and prosper if it does not act to curb GHG emissions in compliance with local, regional, and global environmental laws and regulations. Although any local or regional regulations would only partially impact many international flights (which contribute to approximately two thirds of all aviation GHG emissions), due to lack of jurisdiction over international waters, they still may play a role in affecting industry action.

At the 3rd Aviation and Environment Summit in Geneva, Switzerland, on April 22, 2008, the aviation industry recognized that doing nothing was not an option. A commitment was made to Action on Climate Change by industry leaders and trade associations (**Figure 1**). This action led to an International Aviation Transport Association (IATA) agreement, in 2009.

This agreement, ahead of the UN Climate Conference in Copenhagen, 2009, announced the aviation industry's commitment to a global approach to mitigate aviation GHG emissions, adopting three high-level goals:

1. An average 1.5% fuel efficiency improvement per year from 2009 to 2020
2. Cap net aviation CO_2 emissions from 2020 onward (carbon-neutral growth)
3. A reduction in net aviation CO_2 emissions by 50% by 2050, relative to 2005 levels

These collective goals were endorsed by all participants (airlines, manufacturers, airports, and air navigation service providers) through the International Civil Aviation Organization (ICAO) in September 2009 [3]. A road map for the reduction of GHG emissions was published and is shown in **Figure 2**.

The top line of Figure 2 indicates what would happen if no improvements are made, and the industry continues to grow as forecasted. Next is the potential impact of

FIGURE 1 Aviation industry commitment to action on climate change [2].

Aviation Industry Commitment to Action on Climate Change

As leaders of the aviation industry, we recognise our environmental responsibilities and agree on the need to:

- build on the strong track record of technological progress and innovation that has made our industry the safest and most efficient transport mode; and
- accelerate action to mitigate our environmental impact, especially in respect to climate change while preserving our driving role in the sustainable development of our global society.

Therefore, we, the undersigned aviation industry companies and organisations declare that we are committed to a pathway to carbon-neutral growth and aspire to a carbon-free future.

To this end, in line with the four-pillar strategy unanimously endorsed at the 2007 ICAO Assembly, we will:

1. push forward the development and implementation of new technologies, including cleaner fuels;
2. further optimise the fuel efficiency of our fleet and the way we fly aircraft and manage ground operations;
3. improve air routes, air traffic management and airport infrastructure; and
4. implement positive economic instruments to achieve greenhouse gas reductions wherever they are cost-effective.

We urge all governments to participate in these efforts by:

1. supporting and co-financing appropriate research and development in the pursuit of greener technological breakthroughs;
2. taking urgent measures to improve airspace design including civil/military allocation, air traffic management infrastructure and procedures for approving needed airport development; and
3. developing and implementing a global, equitable and stable emissions management framework for aviation through ICAO, in line with the United Nations roadmap agreed in Bali in December 2007.

Our efforts and commitment to work in partnership with governments, other industries and representatives of civil society will provide meaningful benefits on tackling climate change and other environmental challenges.

We strongly encourage others to join us in this endeavour.

innovation and technology improvements in aircraft design, how aircraft are operated, and the infrastructure. Clearly these efforts alone are not sufficient to meet the high-level goals already mentioned. Additional technology breakthroughs combined with "taking the carbon out of the fuel" are necessary Finally, since these efforts are not expected to be mature and implemented on a large scale in the near or intermediate future, the aviation industry has also developed and agreed voluntary economic measures (a global carbon offset program) that will start in 2020. The goal is to ensure that net GHG emissions from aviation will not increase from 2020, that is, meeting the industry goal of carbon-neutral growth. The next section addresses these improvement areas in order.

FIGURE 2 The IATA road map laid out to achieve high-level GHG emission reduction goals [4].

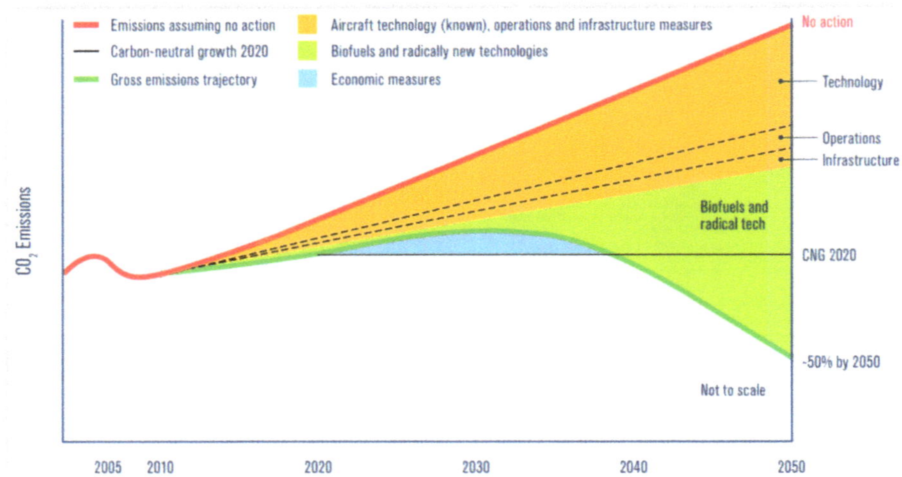

Emission Reduction

Fuel Efficiency and Aircraft Design

New technologies are constantly developed by aerospace companies, and when market conditions are right and promising technologies are mature, new airplane development programs can be launched. When a new airplane is designed and launched, the improvement in fuel efficiency, compared to the model it replaces, is typically at least 20%. Compared to the Boeing 707 at the beginning of the jet aviation age, today's airplane designs have improved fuel efficiency by approximately 70%, a trend shown by **Figure 3**.

Examples of breakthrough innovations are new engine technology, such as high-bypass turbofan engines (e.g., PW1000G, CFM LEAP-1B, PW1500G, Trent 1000, and GP7000) and advanced materials such as carbon fiber composites. Composite materials contribute to an estimated weight reduction of about 20% compared to a more traditional design.

Aerodynamic improvements would include new wing designs made possible because of new materials and engineering models (e.g., winglets, split scimitar winglets, Sharklets), which have claimed fuel improvements up to 4%. Interestingly, these relatively simple improvements can be retrofitted onto many older airplanes. Today most single-aisle aircraft have some form of winglet.

This is the heart of aerospace engineering: designing efficient airplanes. Since fuel, on average, accounts for 20% to 40% of airline operating costs, replacing older models with newer ones is critical for airlines, their operating efficiency and growth (**Figure 4**).

In addition to manufacturers fiercely competing on airplane design, there is also extraordinary collaboration across the aviation industry. One forum for this

FIGURE 3 Historical comparison of aircraft energy intensity [5].

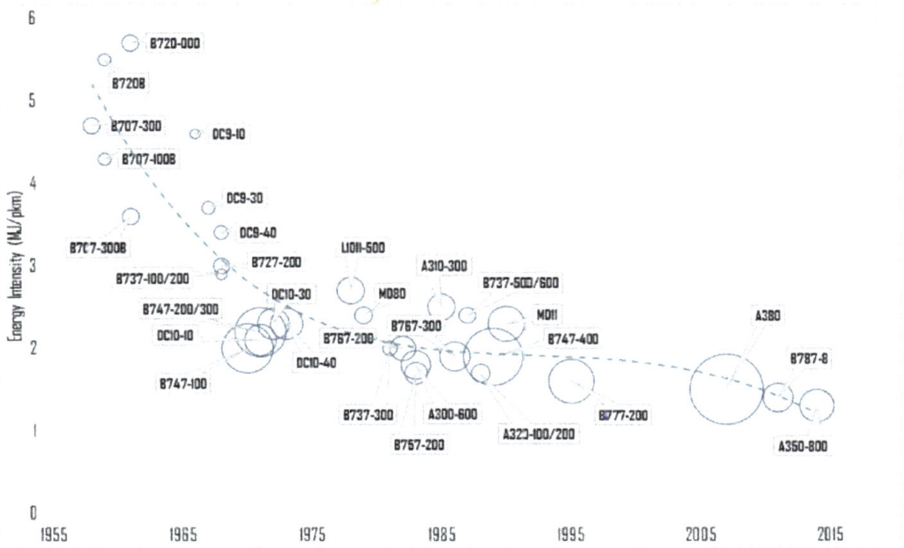

FIGURE 4 Fuel efficiency trends for new commercial jet aircraft 1960–2014 [6].

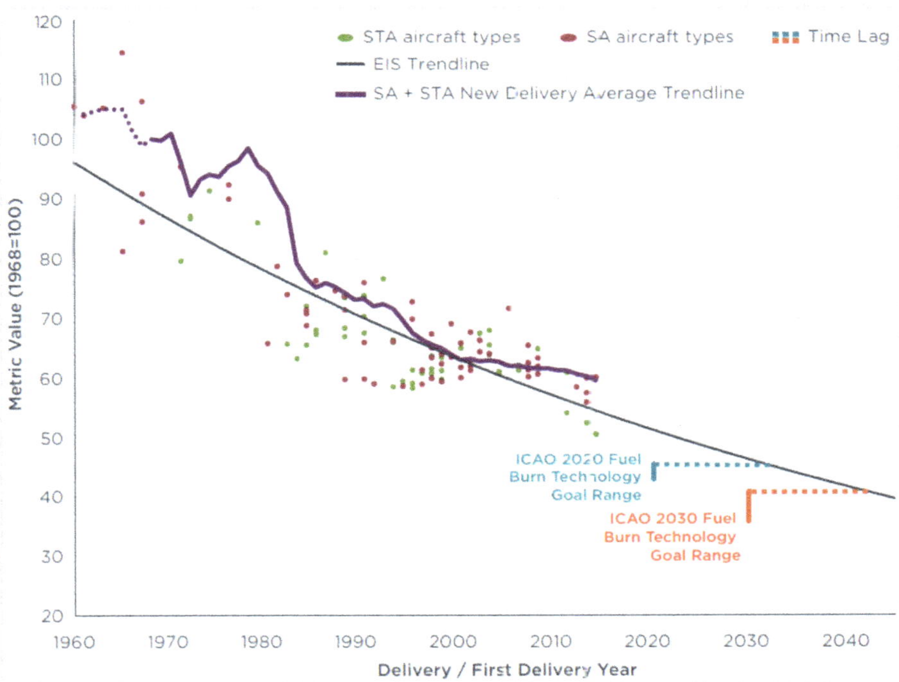

collaboration is ICAO, which sets standards and regulations necessary for aviation safety, security, efficiency, capacity, and environmental protection.

On March 6th, 2017 the ICAO Council adopted a standard for new aircraft CO_2 emissions which represents the world's first global design certification standard governing CO_2 emissions for any industry sector. This standard will apply to new aircraft type designs from 2020, and to aircraft type designs already in-production as of 2023. Those in-production aircraft which by 2028 do not meet the standard will no longer be able to be produced unless their designs are sufficiently modified. (Source: ICAO Newsroom)

Here is yet another example of where the aviation industry, like few other industries, competes intensely on design innovation but at the same time can develop a global standard for fuel efficiency for not only new airplanes models but also for existing aircraft already in production.

Operations and Infrastructure

It is estimated that 10% to 12% of all aviation fuel used is due to inefficiencies in air traffic control (ATC). Most, if not all, commercial flights today are flown using Instrument Flight Rules which use the instruments on the airplane together with radar beacons. In areas of heavy traffic, air traffic controllers direct traffic to use radar in order to track all airplanes in the area. In contrast, private aviation might use Visual Flight Rules where they navigate using visible observation together with dead reckoning, taking account of drift due to wind conditions based on appropriate maps.

Flying a direct route between the city of origin and the destination would not only reduce fuel usage and flight time but would also lead to airlines becoming better able to utilize their aircraft, reducing the cost of getting from point A to point B.

The industry is taking steps to reduce these inefficiencies. The Federal Aviation Administration (FAA) in the United States is leading an effort called NextGen with a focus on modernizing how air traffic is controlled in U.S. airspace. This includes shifting to satellite-based navigation using Global Positioning System satellites from stationary radar beacons. This change enables real-time location, and ground speed information, to determine the most direct route and/or avoid bad weather. It can also optimize the speed of the aircraft to minimize arrival and landing delays and could allow for an efficient continuous descent approach. All these factors benefit pilots using the airspace, the flying public, safety, and the airlines and support progress to reduce the environmental impact of aviation.

A similar effort is ongoing in Europe called SESAR (Single European Sky ATM Research) 2020. This is a research and innovation program to demonstrate the viability of technological and operational solutions in large and operationally integrated environments. Today, European countries typically have their own ATC systems. An integrated system, Single European Sky (SES) is planned to deliver traffic improvements similar to those planned for NextGen. The aim is to triple the airspace capacity, reduce delays, improve safety performance by a factor of 10, reduce environmental impact by 10%, and decrease the cost of air traffic management by 50%.

Similar efforts are occurring elsewhere in the world. ICAO has established a standard for satellite navigation system to set the stage for a future with seamless air traffic management systems providing these improvements across the globe.

Economic Measures

An important short- to medium-term portion of the strategy to meet the target of 50% GHG emission reduction, while air traffic is still growing 5% per year globally, is to use economic measures or carbon offset and reduction schemes. While individual countries can establish economic measures for domestic aviation, international aviation is a different matter and this is where ICAO comes in. Their work is especially important considering that international aviation contributes about two thirds of total emissions from aviation.

If an economic measure for international aviation is to be successful, it must be adopted globally. Therefore, at the October 2016 ICAO 39th Assembly, members adopted resolution 39.2 to establish a Carbon Offset and Reduction Scheme for International Aviation (CORSIA). This is the first global carbon offsetting scheme adopted by the aerospace industry. The resolution calls for a route-based approach where emissions for flights between participating nations are measured. CORSIA is scheduled to go into effect in 2021, using emissions numbers from 2019 and 2020 as a baseline, and has a phased international implementation:

1. 2021–2023 pilot phase
2. 2024–2026 first phase
3. 2027–2035 second phase

The first two steps are voluntary and, as of May 2017, 70 nations have agreed to participate. This accounts for 87.7% of all emissions from international aviation.

So, how does it work?

The baseline for flights between two nations for 2019 and 2020 is calculated and, starting in 2021, the annual GHG emissions are calculated based on fuel consumption reported by the airlines. Any increase in annual GHG emissions is then compensated for with the purchase of market-based carbon credits for projects such as tree planting or replacing fossil fuel energy with renewable sources such as solar and wind. The carbon credits are based on the UN-controlled and verified climate credits program, where 1 credit equals 1 ton of CO_2 mitigated. These offsets are paid by aviation operators. There are many more details of how the scheme works, and further information including monitoring, reporting, and verification can be found in reference [7]. Since many of the projects generating climate credits are thought to be implemented in developing nations, the investments made will also contribute to economic growth in those countries.

Alternative Fuels

As can be seen in the IATA Technology Roadmap (Figure 2), it is not sufficient to only increase fuel efficiency, improve operations and infrastructure, or introduce economic measures for the aviation industry to continue to grow while meeting ambitious pollution abatement targets. The type of fuel used ("take the carbon out of the fuel") also has to be addressed, and it is an essential piece of the long-term strategy for sustainable aviation.

Several alternatives to today's aircraft engines have been developed and tested through demonstration flights. For example, pilots Bertrand Piccard and André Borschberg successfully circumnavigated the globe in 2016 in their solar-powered Solar

Impulse 2 aircraft. This flight included a 5-day flight from Japan to Hawaii in 2015. Although solar power may only provide incremental (hybrid) support power for the foreseeable future, this flight demonstrated promising developing technologies that can be adopted to help the fight against climate change.

In an article in *Aviation Week* from November 5, 2017, it was announced that a start-up company Zunum, with headquarters in Kirkland, Washington, planned to certify and deliver a 12-passenger regional airliner with hybrid-electric propulsion in 2022. This effort has venture-capital backing from both Boeing and JetBlue. The planned range is 700+ miles with a speed of 340 mph based on current battery technology [8].

Similarly, Airbus announced on November 28, 2017, that they are teaming up with Rolls-Royce and Siemens to develop a hybrid-electric flight demonstrator [9].

At least for the foreseeable future, these recent developments can possibly work on short-range regional flights. For longer flights, due to a commercial plane needing a power plant with a high power-to-weight ratio, filling a commercial airplane with efficient batteries would preclude passengers or cargo. Therefore, a critical effort is the development to change the fuel or to "take the carbon out of the fuel" (i.e., biofuel). So what is biofuel, and why is it better in terms of GHG emissions?

Consider **Figure 5** and the various stages that CO_2 can be emitted, and absorbed, in the overall life cycle. Biofuel uses several types of plants or biomass or feedstock as its raw material. This biomass absorbs carbon dioxide from the atmosphere as it is grown, which is then released back into the atmosphere when the fuel is burnt in aircraft engines. Today's jet fuel is based on oil extracted from the ground, which results in no carbon absorption, while biomass acts as a carbon sink, removing CO_2 from the atmosphere.

For biofuels to be an economically viable solution, they must be able to be used as a "drop-in" replacement. If using aviation biofuel, instead of or mixed with standard jet fuel or Jet A, would require a modification to or replacement of the aircraft engines used on 20,000 in-service commercial aircraft, biofuel will not be commercially viable until the cost of oil reaches a high enough level or biofuel becomes inexpensive in comparison. Therefore, the global standard organization ASTM International developed a standard for aviation biofuel which defines properties that would allow it to be used as a replacement for, or mixed with, Jet A. ASTM included this requirement in its global standard for aviation fuel, D1655. This standard defines the characteristics of the fuel to ensure safe operations and is approved by FAA.

On February 24, 2008, Virgin Atlantic flew a Boeing 747-400 from London to Amsterdam using a 20% blend of biofuel in one of its four General Electric CF6 engines [11], in the first known flight to use biofuel. As the second airline to test biofuel, Air New Zealand [12] conducted a 2-hour flight with a Boeing 747 using a 50% blend for one of its Rolls-Royce engines. The engine stood up to use at high altitude, and in other demanding conditions, with no adverse effect on the engine or its performance. Between 2009 and 2013, over 20 additional demonstration flights happened using varying mixtures, feedstock, and engines. One notable demonstration flight was done by a U.S. Navy F/A-18 Super Hornet, aka "Green Hornet," in April 2010, with the aircraft flying as expected, with no degradation of capability.

The first commercial flight using biofuel with paying passengers on board took place on June 30, 2011. A KLM Boeing 737-800 carried 171 passengers from Amsterdam

FIGURE 5 Fossil fuel and biofuel life cycle [10].

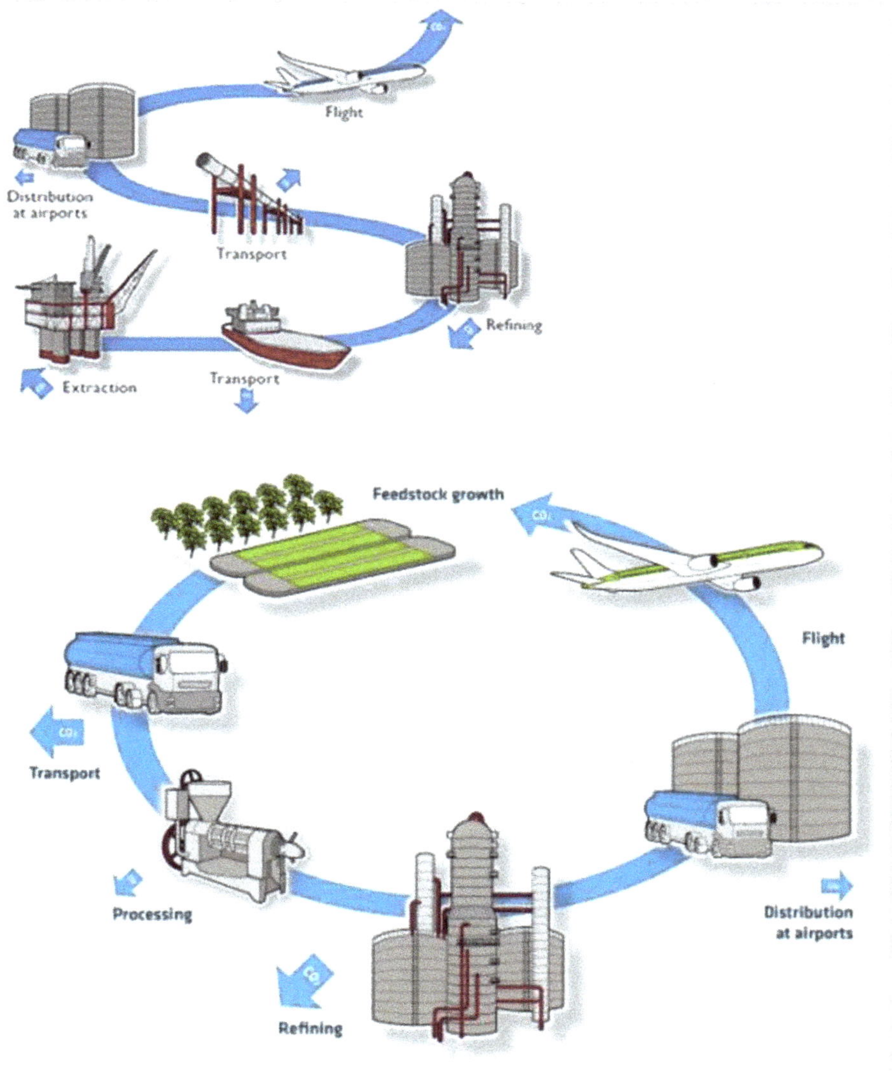

to Paris. Since then, many commercial flights have taken place using various blends, feedstocks, aircraft, and engines. On June 1, 2017, Airbus announced that it has become the first aircraft manufacturer to offer its customers the option of delivering new jets using a blend of sustainable jet fuel.

While the examples above demonstrate that using biofuel in aviation is technically feasible, there are still many challenges to be overcome. One of the key issues, apart from availability (more about that later), is that biomass needs space to grow. With over 7 billion people living on the planet, and an expectation of 9 billion by 2050, the question may be one of feeding people or growing plants to be converted into biofuel for aviation. Therefore SAFUG, the Sustainable Aviation Fuel Users Group [13], with its membership of leading airlines representing about one third of all commercial aviation, has made a pledge to commit to sustainable options for biofuel. Their pledge is that, at a minimum, the following considerations should be addressed by verifiable means: jet fuel plant sources should be developed in a manner which is noncompetitive with food and where biodiversity impacts are minimized; in addition, the cultivation of those plant sources should not jeopardize drinking water supplies [14].

Examples of second-generation feedstock for biofuels are:

- *Camelina* is primarily an energy crop, used as feedstock to produce renewable fuels. Note that the leftover solid "meal" from the oil extraction process can also be used as a feed supplement for poultry and livestock, that is, nothing goes to waste. Camelina can be grown as part of a regular crop rotation program with wheat and other cereal crops, so it doesn't compete with food production. It is also a water-efficient crop.
- *Halophytes* are salt marsh grasses and other saline habitat species that grow in areas where no useful food crop can grow, making it a perfect crop that doesn't affect food production or fresh water supply.
- *Jatropha* is another energy plant that prospers on land where it is difficult or impossible to grow food crops, e.g., arid land. The plant produces toxic, inedible seeds with high oil content that can be used for biomass.
- The most promising feedstock for aviation biofuel is *algae*. This is the least developed feedstock but the one with the most promise. In addition to being able to grow in places typically not suitable for food production, given a certain surface area, algae can produce up to 15 times more feedstock than the three options mentioned above.

There is further development to use waste products such as food, forestry, and agricultural waste products as feedstock for renewable jet fuel. These additional feedstocks would use material that would otherwise wind up in landfills. Imagine a KLM flight out of Amsterdam where "fritjes met sla" are served while the aircraft is powered by renewable jet fuel based on used cooking oil.

Biofuel has proven to be a safe and environmentally sound complement and alternative to Jet A. A significant remaining challenge is to commercialize and scale the development to make sustainable biofuel affordable and available at the quantities needed. This requires a concerted effort involving airlines, aircraft manufacturers and governments, refineries, producers of feedstock, airports, and certification bodies to work together.

Aircraft Noise

For the aviation industry to continue to grow and double in size in less than 20 years, the infrastructure needs to keep pace. Particularly, airports must be able to handle additional traffic. A major concern is the noise experienced by those living along the flight paths of departing and arriving aircraft. Let's look at Heathrow, the second busiest airport in the world for international passenger traffic, as an example. The airport was established in

1929, when the surrounding area was mostly farmland and market gardens. During the war, the area was developed to be a major air base for military aircraft. However, after the war, it was designated as an airport for civil aviation. It was called London Airport when it opened in 1946 and it wasn't until 1966 it was renamed Heathrow. Since then, the city of London continued to grow and the area around the airport was soon developed as residential. Air traffic noise, from takeoffs and landings, continued to increase and it became an important concern for those living nearby. In order to accommodate the increase in traffic, a third runway at Heathrow was first studied by the Roskill Commission in 1968. Plans have since been made but never materialized due to economic and financial reasons. In addition to concerns about increasing global GHG emissions, several hundred thousand people in the flight paths would be impacted by an increase in aircraft noise. In June 2017, the third runway proposal was again endorsed by the transportation minister and is now proceeding with the planning and permitting phase. The House of Commons voted 415-119 in favor of a third runway on June 25th, 2018. The decision to proceed is challenged in court and a final decision to proceed with the expansion is not expected until after all judicial reviews are completed.

The same is true for many cities where aviation development occurred early in the era of commercial aviation, e.g., Reagan Airport, Washington DC; Orly Airport, Paris; and Haneda Airport, Tokyo.

Newer airports are located further away from city centers (e.g., Narita Airport, Tokyo) or some even on artificial islands (e.g., Chek Lap Kok, Hong Kong, or Kansai Airport, Osaka). The cost of building new can be very significant, e.g., the new Beijing Airport planned to open in 2019 at a cost of $12B. The new airport in Beijing will be 42 miles south of the capital [15].

Locating airports further away from population centers means that, where train travel is an option, it becomes a serious contender for short haul "door-to-door" travel.

The aviation industry has collaborated for many years to address the issue of noise. ICAO has adopted a strategy shown in Figure 6 for aircraft noise.

Taking each step of their balanced approach in order, starting with step 1, reduction of noise at source:

ICAO developed Chapter 1, their first noise standard, in 1973. The latest standard, Chapter 14, was adopted in 2014, and it is applicable to new airplane types submitted for certification on or after December 31, 2017. These standards have been adopted by national certification bodies such as FAA and EASA. As a result of the implementation of the Chapter 14 noise standard, it is expected that more than 1 million people would be removed from a DNL (day–night average sound level) of 55dB between 2020 and 2036 (Figure 7).

FIGURE 6 ICAO aircraft noise strategy [16].

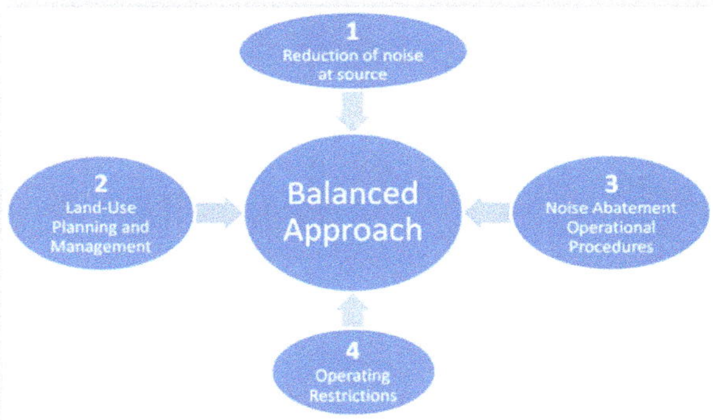

The four principal elements of the Balanced Approach to Aircraft Noise Management

FIGURE 7 Progression of ICAO noise standards.

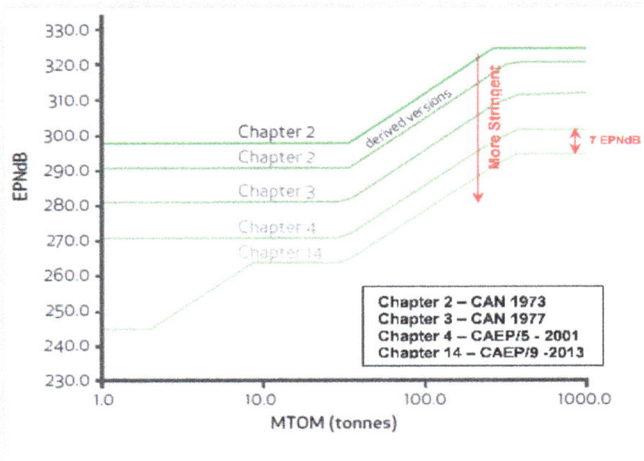

The progression of the ICAO Noise Standard

FIGURE 8 Nacelle design for reduced noise.

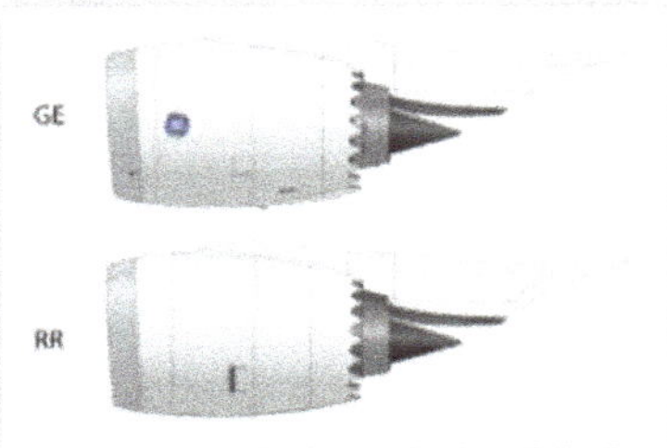

Addressing the strict noise requirements at their source is accomplished through innovative propulsion and aerodynamic designs, such as nacelles with chevrons (Figure 8), which can be found on planes like the Boeing 787.

The second step of the strategy has to do with land use, planning and management. This means to ensure that activities developed nearby airports are compatible with aviation. The main goal is to minimize the population affected by aircraft noise by introducing land-use zoning around airports. ICAO has issued a manual to help assess impact [17].

The third part of their strategy is the day-to-day operational procedures at airports that are both safe and cost-effective. This includes determining the best procedure for takeoffs and landings considering weather and traffic, minimizing noise and maintaining safety of flight, while determining climb rates (flight paths) that are also fuel efficient. For airports with several runways, the selection of runway could also have an impact.

Finally, operational restrictions are measures taken by the airports to limit the traffic allowed at different times of the day. In general, if airports allow nighttime operations, they would prefer most of their cargo traffic later at night and passenger traffic during the day. However, this presents a problem since cargo planes are typically older airplanes (many are converted from old passenger airplanes) and therefore noisier. Therefore, a case can be made for using newer airplanes as cargo planes to service markets around busy airports in heavily populated areas. Not only are the new airplanes more fuel efficient, but they can operate during the off-hours.

Conclusion

The aviation industry has taken or plans to take action to address key environmental issues such as contributions to climate change through greenhouse gas emissions and noise impact from aircraft near airports. As shown here, there is a common approach to solving these critical problems. Feasible solutions all require collaboration among stakeholders such as airlines, aircraft manufacturers, airport operators, and national certification bodies to first agree and recognize the problem and then to have a global strategy to address the issues. One example is the IATA road map with an aim to reduce GHG emission by 50% by 2050 where individual parts of the system have to contribute, or addressing the issue of aircraft noise as traffic volume increases through not only addressing the design of the aircraft but also optimizing flight operations around the airports, and even providing guidelines for how to zone the areas around the airports to minimize the impact on people. It is important to note that this collaboration can take place in spite of the fierce competition between airlines and aircraft manufacturers. This could not have been accomplished without established platforms for collaborations such as ICAO and IATA. Finally, there is a very clear understanding that without solving these problems, the aviation industry will not be able to continue to prosper in the long run.

References

1. UN IPCC 2007 Report, http://www.ipcc.ch/publications_and_data/publications_ipcc_fourth_assessment_report_synthesis_report.htm.
2. ATAG, http://www.atag.org/our-activities/climate-change.html.
3. Group on International Aviation and Climate Change (GIACC) Report No 1, https://www.icao.int/environmental-protection/GIACC/GiaccReport_Final_en.pdf.
4. IATA Technology Roadmap 2013, https://www.iata.org/whatwedo/environment/.../technology-roadmap-2013.pdf.
5. Hofstra University, "The Geography of Transportation Systems," https://people.hofstra.edu/geotrans/.
6. The International Council on Clean Transportation, "Fuel Efficiency Trends for New Commercial Jet Aircraft: 1960-2014, Aug 2015, Fig. ES-2, http://www.google.co.uk/url?sa=t&rct=j&q=&esrc=s&source=web&cd=1&ved=0ahUKEwjk04zer4zYAhViLcAKHX1GDh8QFggpMAA&url=http%3A%2F%2Fwww.theicct.org%2Fsites%2Fdefault%2Ffiles%2Fpublications%2FICCT_Aircraft-FE-Trends_20150902.pdf&usg=AOvVaw382jiykX05yxZ8-w9P5Qxi.
7. www.icao.int.
8. http://aviationweek.com/new-civil-aircraft/boeing-backed-zunum-s-first-aircraft-be-12-seat-commuter?NL=AW-05&Issue=AW-05_20171006_AW-05_40&sfvc4enews=42&cl=article_1&utm_rid=CPEN1000002536587&utm_campaign=12038&utm_medium=email&elq2=380a64ad64fe40858c99701923c02140.
9. http://www.airbus.com/newsroom/press-releases/en/2017/11/airbus--rolls-royce--and-siemens-team-up-for-electric-future-par.html.
10. ATAG, "A Beginners Guide to Aviation Biofuel," 2011, http://www.google.co.uk/url?sa=t&rct=j&q=&esrc=s&source=web&cd=1&ved=0ahUKEwi9wsXGxYzYAhUIJMAKHb94Bw8QFggpMAA&url=http%3A%2F%2Fwww.atag.org%2Fcomponent%2Fdownloads%2Fdownloads%2F60.html&usg=AOvVaw3N7GPsTGDlNB__baqdhoN_.
11. http://news.bbc.co.uk/1/hi/7261214.stm.
12. https://www.greentechmedia.com/articles/read/biofuel-powers-air-new-zealand-test-flight-5436#gs.ZADHaBg.
13. SAFUG, http://www.safug.org/.
14. http://www.safug.org/safug-pledge/.
15. http://www.businessinsider.com/beijings-daxing-airport-biggest-world-2017-10/#beijing-broke-ground-on-the-project-in-2014-under-president-xi-jinping-1.
16. https://www.icao.int/environmental-protection/Pages/noise.aspx.
17. https://store.icao.int/index.php/amendment-no-1-to-doc-9184-part-2-edition-no-3-dated-27-3-09.html.

CHAPTER 11

Standards: Bedrock to Build On

David Alexander
SAE International

> *Today's standardization...is the necessary foundation on which tomorrow's improvements will be based. If you think "standardization" as the best you know today, but which is to be improved tomorrow – you get somewhere. But if you think of standards as confining, then progress stops.*
>
> Henry Ford, 1926

Introduction

What is developed in committee, yet a powerful tool for innovation?
What is "as civil as possible, as military as necessary"?
What is developed by industry, yet relied upon by government?

Standardization has, almost from its earliest moments in the 20th century, been a bedrock for progress of aviation. We will begin by looking at the history of standards in aviation and the origins of some of the main industry standards development organizations, including the huge role of military aerospace in standards. Subsequently this chapter, incorporating the book's areas of focus, will explore some of the main areas in which standards support aerospace: assuring supply chain through product and process conformity and quality management, defining end-user requirements through procurement, providing means of compliance for product certification in the regulatory environment, supporting operations and maintenance, and giving a common baseline for innovation.

But first, we need to review what we mean when we talk about standards. A useful definition for the purposes of this chapter: Standardization is the process of implementing and developing technical standards based on the consensus of different parties that include firms, users, interest groups, standards organizations, and governments [1].

There are a number of instances where the noun standard or standardization is used in aviation: for example, to describe the regulations themselves (the Federal Aviation Administration's (FAA) use of the term in "airworthiness standards") [2], the term standardization being used by the European Aviation Safety Agency (EASA) to describe the adherence to regulatory standards across almost 50 European Union (EU) member states and neighboring countries, or in the International Civil Aviation Organization's (ICAO) Standards and Recommended Practices.

However, in this chapter we will focus on industry consensus standards: those which have been drafted and approved by multiple parties whether in open fora or in fixed membership consortia, with the resultant content able to be used by various organizations.

Industry standards are used for many reasons, including enhancing safety and providing guidance for compliance to certification; technical reasons such as interoperability between different systems, common interfaces, uniform testing, tooling or optimized production processes; and factors such as facilitating trade, connecting and opening global markets, and reducing costs and increasing supply-chain efficiencies through common parts. And they have carried aviation since its early days.

The History of Standards and the Aerospace Industry

Standards have been assisting human development, trade, and interactions since recorded history began: examples include weight and measurement standards for trade, and standardized length measurements for town planning in the Bronze Age Indus Valley civilizations [3].

More recently, standardization played a key role in the 1760–1840 Industrial Revolution through the standardization of screw thread sizes and the interchangeability of nuts and bolts [4] and later on by Joseph Whitworth's screw threads which were adopted nationally in Britain in 1841 as well as in other countries, and were referred to as the "British Standards Whitworth"—an institution before the institution.

Organized standardization on a national level came in 1901 with the establishment of the Engineering Standards Committee, the forerunner of the British Standards Institute, and then Deutsches Institut für Normung in 1917, and the American National Standards Institute and the French Commission Permanente de Standardisation in 1918.

In the meantime, mobility industry pioneers were beginning to work together on standards. With the emergence of the automobile industry in the early 20th century, there became a recognition of the value of adhering to common practices and for cooperation and standardization of certain methods and parts.

In 1902, Peter Heldt of *The Horseless Age*, supported by Horace Swetland of *The Automobile*, wrote in an editorial, "Now there is a noticeable tendency for automobile manufacturers to follow certain accepted lines of construction, technical questions constantly arise which seek solution from the cooperation of the technical men connected with the industry. These questions could best be dealt with by a technical society. The field of activity for this society would be the purely technical side of automobiles" [5].

Likewise, the aviation pioneers, in the decade after the first powered flight by the Wright brothers, were starting to produce increasing numbers of aircraft which were taking part in sustained operations. While certain commercial aviation services (such as mail, passenger, and freight) were beginning, it was in large part military applications that really sustained the industry: in the pre-war buildup for reconnaissance and photography and then in World War I itself for aerial superiority. Early aircraft companies including Sopwith, De Havilland, Junkers, Curtiss, Bleriot, and Breguet turned out numerous models involved in dogfights and other supporting roles.

This transition from one-off models to large-scale production, allied to the need to produce aircraft quickly to support the war effort, led to the recognition of standards. In 1916, the Society of Automobile Engineers (SAE International)—originally established in 1905—expanded its remit to include aeronautics as well as tractors: SAE member Elmer Sperry actually created the term "automotive" from Greek autos (self) and Latin motivus (of motion) origins to represent any form of self-powered vehicle. The Society of Automobile Engineers became the Society of Automotive Engineers [6]: Henry Ford's quote at the beginning of this chapter is as relevant to aviation as to automobiles, and arguably any other industry sector.

Articles on aeronautical technologies had been published in SAE Journals and Bulletins since 1917, with the first SAE aerospace standard, a common thread specification for an aircraft engine spark plug, published in an SAE Bulletin in January 1917.

The Aerospace Standards Landscape: Early Standards Development Organizations

Other landmarks for aerospace-specific standardization came in addressing the need for safe and reliable navigation through radio equipment and the communications with ground stations and waypoints: Aeronautical Information Incorporated (ARINC) was established in 1929 and chartered by the U.S. Federal Radio Commission to coordinate U.S. nongovernment radio communications for the airline industry. Today, ARINC Industry Activities manages hundreds of standards set by the airlines and industry for electronic, avionic, and other airborne equipment, in many cases carrying the ARINC standard number (such as 429) as an informal brand—and which are applied around the world.

Elsewhere, RTCA, chartered by the FAA, was founded as the Radio Technical Commission for Aeronautics in 1935 and today RTCA standards "serve as the basis for government certification of equipment used by the thousands of aircraft flying daily through the world's airspace" [7]. In many cases partnering with RTCA and SAE International, EUROCAE was established in 1963 in Lucerne as the European Organisation for Civil Aviation Electronics (later the European Organisation for Civil Aviation Equipment). In 1966 it was formally recognized by ECAC as a recommended source for specifications to be used by aviation authorities. EUROCAE today maintains 40 active working groups and has provided a leadership role in European aviation standards coordination, most notably for air traffic management (ATM), unmanned aircraft systems (UAS), and cyber security. Also in Europe, the aerospace industry trade association, AECMA (now ASD), established industry-driven standards activities in 1973 covering standard parts and materials, later broadening to encompass business process and logistics protocols. In the United States, ASD's counterpart Aerospace Industries Association, or AIA, was founded in 1919 and maintains a large portfolio of globally ubiquitous parts standards (fasteners) as well as partnering with ASD on the business and logistics standards.

Value and Drivers

Standards therefore have been extremely important to the progress and maturity of the aerospace industry, and the development and maintenance of a standard do require investment of time and resources on behalf of its authors before it enters common usage. This investment is made worthwhile through value and drivers found in safety, design, production, operation, and maintenance of aircraft.

Standards may have a wide range of uses, from detailed mechanical standards to process standards and recommended practices. The intended use of a standard is sometimes represented through a signifier within its title (such as SAE International's Aerospace Standard (AS), Aerospace Material Specification (AMS), Aerospace Recommended Practice (ARP), and Aerospace Information Report (AIR) designations). However, the intended use of a document may be noted in the document's title, such as American Society for Testing and Materials' (ASTM) "Standard Practice for ..." or "Standard Specification for..." or EUROCAE's "Minimum Operational Performance Standards" (MOPS), "Minimum Aviation System Performance Standards" (MASPS), and "Interoperability Standards" (Interop). Understanding this intended use can be particularly important to ensure that guidance is used in deploying the standard, particularly when addressing a compliance context.

Two elements in particular describe the importance of standards defining common ground in aviation: the criticality of safety and the global nature of flying.

Safety: Particularly in civil (transport) aviation, travelers do not typically choose which aircraft brand, nor model, they fly or specify which engines will power the flight or whose avionics systems will provide for the electronic controls, connections, and communications networks. And the travelers are very much passengers, with strongly locked doors separating them from the piloting of the aircraft. As a result, once a product has been certified, the aerospace industry cannot compete on safety nor on the quality conformance of product. An accident will affect the industry, and end-customer (passenger) confidence, of aviation as a mode of transport. Commercial transport aircraft also carry inherently high-risk factor: a catastrophic error that leads to an accident may well kill hundreds of passengers instantly. So the aerospace industry must ensure that it works to, and meets, the most stringent standards: which, considering that aviation is one of the very safest modes of transportation, is a testament to the industry's investment in safety [8].

Global nature: While the aerospace industry is growing with new players in "new" and legacy aerospace countries developing aircraft, aerospace still provides a global paradox—while the design of products is concentrated in a relatively small number of countries by a small number of (mostly large) companies, civil transport aircraft are owned, operated, and maintained in every country in the world, with an increasingly global supply chain and manufacturing base. Therefore, standards developed and "owned" by the industry that meet the very specific requirements of aerospace use are necessary: interoperability beyond borders ensuring that flight operations can be carried out across the world, maintenance is performed, and spare parts delivered as soon as possible to enable aircraft availability is critical.

Indeed, standards also play a critical role in supporting the life cycle of an aviation product: the data in the standard supporting a product, system, or piece of equipment will be used not just in the design but also to enable inspection and where necessary maintenance and/or replacement: because aircraft must meet continuing as well as airworthiness requirements, installed and replacement products should also meet the three Fs of product standards throughout their lifetime—fit (the ability of a product to connect with or join to another part or in an assembly), form (characteristics such as materials, weight, size), and function (the ability of the product to perform its intended purpose).

Standards and Conformity

Where we talk of some (industry) standards in the most formal sense, as performance standards or requirements, industry standards at the point of publication are voluntary and non-binding. There are however a number of ways in which adherence to these documents can become mandatory or quasi-mandatory within the contextual limits. This may be done contractually, when a standard is called out as a design, process, or other condition in a commercial contract, often through requirements flowed down a supply chain. In a similar way, a customer may specify that a conformity or qualification requirement be satisfied, for example, accreditation to a process requirement which references a specification in an auditable checklist, or the qualification of a product to the criteria set out in a standard.

In some cases, this conformity is actualized before the user takes the part rather than the product being tested against the criteria of the requirements (e.g., in a standard) by the user. Where qualification of the process or product is required up-front, an independent qualification program may be used. There are a number of examples of this: for special processes, the National Aerospace and Defense Contractors Accreditation (Nadcap) program administered by the Performance Review Institute (PRI) uses expert contracted auditors to assess process compliance against criteria—usually based on prime requirements or standards such as AMS2750 for pyrometry or International Organization for Standardization (ISO) and ASTM test methods—set by original equipment manufacturers (OEMs) who then take the final decision on accreditation. For products, notably fasteners, connectors, elastomer seals, and other parts, and components, the Aerospace Standards and Part Qualification Program under SAE Industry Technologies Consortia (ITC), the ASD-CERT program, and the Performance Review Institute Qualified Product List (PRI QPL) use industry expert and user resources to assess and qualify products against standards criteria, with the approved products being listed in a database known as a QPL which can be used for procurement purposes. Through such schemes, conformance to quality and form, fit and function, is assured before delivery.

Quality Management Systems

Strongly connected to conformity, one of the most easily identifiable uses of standards is in ensuring product quality. Many who may not (at least knowingly) work with standards may be familiar with the ISO9001 Quality Management Systems standards which can be applied to many manufacturing and service sectors. However, as noted in the introduction, because of the aerospace industry's very stringent safety requirements, quality takes on another level of criticality.

In 1998, the aerospace industry formed an association, the International Aerospace Quality Group (IAQG), to develop common quality management system standards, based on the principles of ISO9001 but addressing the specific needs of the aerospace and defense sector. The main standard, AS9100 "Quality Systems - Aerospace - Model for Quality Assurance in Design, Development, Production, Installation and Servicing," has been backed by a multifaceted and robust certification scheme known as Industry Controlled Other Party (ICOP) which involves accreditation bodies, certification bodies, auditors, and of course the quality management community in industry.

In 2018, over 22,200 sites around the world are certified to the 9100 series of standards, all available publicly for procurement and conformance purposes in the Online Aerospace Supplier Information System (OASIS) database. Following outreach efforts, civil regulators and military user organizations are starting to recognize the standards and the ICOP scheme as assurance for the industry.

Standards and Innovation

Upstream from quality and conformity is research, technology, and an ongoing debate in the context of standardization: the role of standards in supporting innovation.

Standards are a prerequisite for certain technology advancements—a new piece of equipment installed on an aircraft must fit into a bigger system, and ensure that it meets environmental requirements, etc. On a much larger scale, standards ensuring interoperability in new air traffic management systems for aircraft to operate in multiple airspace jurisdictions during the journey from takeoff, through the flight, to landing are needed. But it is sometimes less clear where the role of standards begins and ends with innovation.

An individual or organization wishing to "rip up the rulebook," to define a truly disruptive model, service, or product, may find that adherence to existing standards is unhelpful and that the context in which those standards operate is too restrictive to truly creative design. However, as a pioneering design progresses, it is necessary to consider the implementation of that particular innovation in its context. And in this case, standards can be a platform or a framework to guide the innovation.

The "sweet spot" for innovation to benefit from industry consensus varies from case to case, and will depend heavily on the technology itself but will always include the point at which industry—design authorities, partners, competitors, and supply chains—is willing to share in order to collectively compile certain aspects of a future development. Intellectual Property (IP) is an extremely valuable asset for any organization, and the IP holders of innovative products and approaches must be prepared to share in order for a new technology to be described in a standard or recommended practice. This in turn implies that industry recognizes that such collaboration and standardization will advance a new technology, concept, or product to market through ideation, design, certification, manufacturing, and operation more quickly than each organization would do independently.

An example of where the industry consensus process has used "standards" for an innovative technology is in the real-time monitoring of flaws in aircraft structures outside visual maintenance inspections. In 2013, after 5 years of work on a groundbreaking project, the Aerospace Industry Steering Committee on Structural Health Monitoring published ARP6461, "Guidelines for Implementation of Structural Health Monitoring on Fixed Wing Aircraft." While a recommended practice describing a range of approaches rather than a standard, because of the partnership between industry and the regulatory agencies in the development of the content, it provides guidance commonly agreed to by the industry and is understood by the applicant and certification agency which may simplify the certification process.

Collaborative Endeavors for Standards Development

As innovation flourishes to support large-scale or aviation system-level needs through standardization, certain responsible bodies are developing coordination groups to ensure that (a) standard(s) can be generated at the pace in which technology, usage, or regulatory compliance requires and (b) the correct standards are generated for the need. This may be for technologies and concepts such as next-generation air-traffic management systems and the operation of UAS or drones in the airspace or for new technology or manufacturing concepts, such as additive layer manufacturing (including 3D printing) and digital manufacturing. These groups, typically comprising of the Standards Development Organisations (SDOs), key industry, government and user groups, and relevant research organizations, work collectively on a national, regional, or general industry basis and may utilize a framework and road map to monitor progress.

Examples of this include:

- The European ATM Standards Coordination Group which coordinates the timely development of standards to support the modernization of Europe's ATM infrastructure for the Single European Sky [9]
- The America Makes and ANSI Additive Manufacturing Standardization Collaborative which facilitates the growth of the additive manufacturing industry through the development of industry standards for additive manufacturing [10]
- The European UAS Standards Coordination Group which maintains a Rolling Development Plan of standards from various SDOs involved in generating standards supporting UAS and drones—as well as needs from the regulatory and operational community
- A counterpart in the United States, the UAS Standardization Coordination, providing coordination to avoid duplication of effort and to achieve a coherent set of standards [11]

Dual-Use Standards and MilSpec Reform

While this book focuses on civil aviation, it is necessary to continue to highlight, as we did at the beginning of this chapter, the crucial role played by military aviation in developing, promulgating, and utilizing standards that the entire aviation system has benefitted from.

Indeed, it is estimated that 3,400 SAE standards alone have been adopted by the U.S. Department of Defense, going back to Liberty Trucks in WWI: similarly in the United Kingdom, Defence Standards (DEFSTANs) provided the basis for many standards used in the commercial aerospace industry, and likewise in France through AIR standards, Germany with Luftstelle Norm standards, and in other countries with a significant defense capability and industry.

In the United States, a series of standardization mechanisms where developed: Defense Handbooks (MIL-HBK) which compile information and guidance, Defense Specifications (MIL-SPEC) which detail essential technical requirements, Defense Standards (MIL-STD) which provide uniform engineering and technical requirements, Performance Requirements (MIL-PRF) which defines functional and environmental characteristics, and Detail Specifications (MIL-DTL) specifying design requirements [12]. These standards came to define many of the aerospace products in the 20th century, both defense and commercial.

However, while military standards, specifications, and handbooks have remained the bedrock of many recent and current aerospace programs, newer standards activities have in many cases been taken on by the civil aviation community. And arguably a paradigm shift has occurred over the past 20–25 years, as military organizations have reduced their in-house standardization generation and work at the detail level in order to focus on standardization management: in doing so, transferring the responsibility for that detailed work on standards to nongovernmental organizations.

The most famous example of this came in 1994, when the U.S. Department of Defense (DoD) began what has been commonly known as MilSpec reform, or the Perry Reform, after the then Secretary of Defense, Dr. William J. Perry [13].

Almost 20,000 documents, estimated to be a half of the overall U.S. DoD standardization portfolio, were removed from government control between 1994 and 2001 [14], with many being transferred to industry standards bodies such as SAE International, ASTM International, and AIA.

This principle has been introduced at North Atlantic Treaty Organization (NATO) through the Civil Standards Campaign to increase the reliance on Commercial Off-the-Shelf products and industry-maintained standards for NATO interoperability for nonoperational reasons. And to facilitate supply chain efficiencies between civil and military industry and promote interoperability, dual-use standards, those intended to be used for both military and civil purposes, are increasingly common.

In the area of air traffic management, interoperability standards support opportunities to facilitate the shared use of airspace between civil and military traffic. In Europe, air traffic control (ATC) providers on a national and those on a pan-European level such as European Organisation for the Safety of Air Navigation (EUROCONTROL) are collaborating with industry through industry standards bodies such as EUROCAE. This includes the integration of military UAS into civilian airspace and the use of military airspace by commercial traffic afforded from a reduction of military traffic [15].

The "civilianization" of these former military specifications and dual use is a good example of the harmony through standards of the civil and defense aerospace markets.

Standards in Civil Aviation Regulations

Returning to the civil arena, standards play an increasingly important role in supporting the civil regulatory process, particularly in certification. Indeed, industry standards support the positive trend of regulatory harmonization between aviation authorities—such as those for Europe (EASA), United States (FAA), Canada (Transport Canada Civil Aviation (TCCA)), Brazil (Agência Nacional de Aviação Civil (ANAC)), Japan (Japan Civil Aviation Bureau (JCAB)), China (Civil Aviation Administration of China (CAAC)), Singapore (Civil Aviation Authority of Singapore (CAAS)), and others—and the international work of the ICAO with the goal of mutual recognition of approvals and global aviation safety. However, it is the widely recognized EASA and FAA from which we will draw examples.

EASA and FAA regulatory material commonly referencing industry standards includes EASA's Certification Specifications (CS), Acceptable Means of Compliance (AMC), and Guidance Material (GM) and in the U.S. Advisory Circulars, which provide guidance for compliance with FAA regulations on airworthiness, pilot certification, operational standards, training standards, etc. and Airworthiness Directives (ADs), which are legally enforceable regulations issued by the FAA in accordance with 14 CFR part 39 to correct an unsafe condition in a product [16].

FIGURE 1 Industry Standards Documents/SAE International.

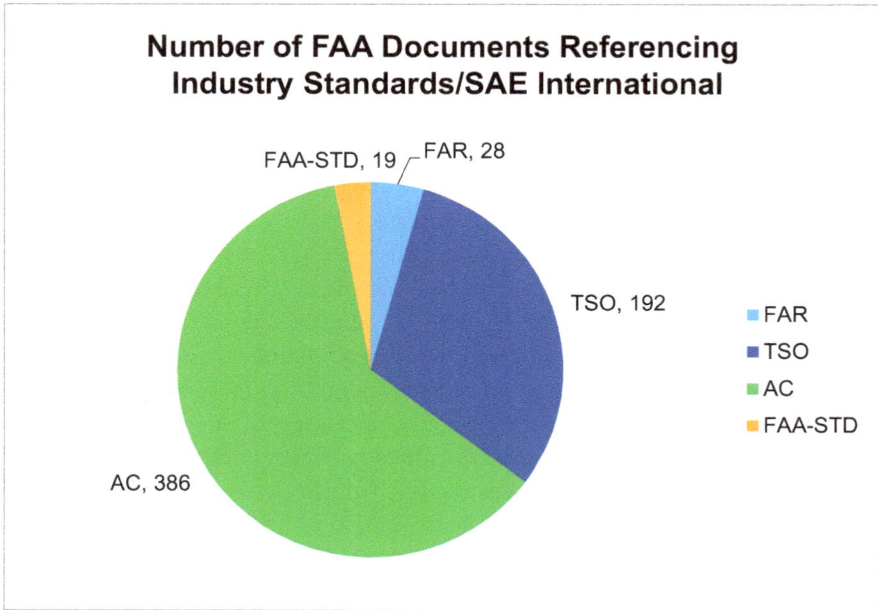

Technical Standards Orders, or TSO (European Technical Standards Orders (ETSO)), set the minimum performance requirements for aviation equipment from safety belts, to landing gear systems, to transponders, to instrument displays and are largely written with reference to an industry standard, from EUROCAE, RTCA, and SAE, as an acceptable means of compliance.

Certain aircraft certification criteria for civil aircraft and products rely heavily on industry standards—CS/Part 25, for example, are the specifications for large airplanes. While many sections provide detailed guidance or requirements for applicants within the regulatory document itself, standards are also referenced. For safety assessment, SAE ARP4754A/ED-79 *Guidelines for Development of Civil Aircraft Systems* "provides a recommended practice for system development assurance" in AC 25. Other examples include ETSO C119d referencing EUROCAE ED-143, *Minimum Operational Performance Standards for Traffic Alert and Collision Avoidance System II (TCAS II)*. CS-25 calls out the application of an aircraft anti-icing fluid compliant to the SAE AMS1428 (types II, III, or IV) specification as an acceptable means of compliance for ensuring wing leading edges, and in AMC 25-11, Electronic Flight Displays refers to SAE AS8055 *Minimum Performance Standard for Airborne Head Up Display*.

Through the standards development process in collaborative stakeholder technical committees or working groups, this provides therefore for a mutually beneficial partnership between industry, the regulatory authorities, and other actors: the certification guidance and specifications benefit from the industry expertise; the industry thus plays a role in the criteria it will ultimately meet.

Furthermore, one benefit of using standards is that where agency-controlled rules and documentation may be less burdensome to change than formal government legislation, so industry standards are more flexible than regulatory and certification documents because of the relative ease of revising or adapting a stand-alone industry document in an standard development organization (SDO) consensus process versus a change in governmental rules. The use of the standards by reference mitigates somewhat the risk of the rules and the industry best practice becoming out of sync owing to different revision processes and cycles.

However, just as overly prescriptive standards may limit innovation, so prescriptive regulation can prohibit industry from finding new solutions and advancing technology. At this point it is worth recalling from the introduction that industry standards are voluntary and, even if included in a regulatory document, are an "acceptable (and not the only) means of compliance." Therefore, the aviation regulatory community is working toward performance-based regulation, which sets higher-level criteria for safety or operation and leaves the applicant latitude to design the product or system to meet those criteria. And standards are an enabler of this performance-based approach by providing non-binding, voluntary guidance.

A logical development, but a significant change, in the regulatory/standardization paradigm was embodied by the Part/CS 23 revision. Part/CS 23, the Certification Specifications Normal, Utility, Aerobatic and Commuter Aeroplanes, underwent revision with an Aviation Rulemaking Committee (ARC) in 2012. The FAA, EASA, and other regulatory partners worked with industry—particularly through industry trade associations such as General Aviation Manufacturers Association (GAMA), ASD, and AIA—to define a new approach.

In this, the regulatory requirements are performance based, with much of the detail that had previously been in the various regulatory instruments instead being documented in industry standards. An ASTM International committee, F-44 on General Aviation, which includes several subcommittees addressing specific subjects such as aeroelasticity, loss of control, and crashworthiness was chartered in 2012. This committee developed industry standards supporting the Part/CS 23 revision, with other existing and new supporting industry standards and guidance able to be referenced in turn.

This methodology facilitates performance-based regulation, reducing the regulatory requirements to high-level safety objectives. From the FAA's final rule summary: Part 23 would be amended by "the airworthiness standards replacing current prescriptive design requirements" [17], and this approach may be in consideration for future regulatory revisions.

Standards Use and Content Consumption: Past, Present, and Future

Standards therefore may be used for reference, as consultative material, when working on a product design or modification. At the parts and materials level, the design engineer (or process engineer) will call out a standard to be used in the product. The implementer, whether from the OEM company or a subcontractor, will then design, manufacture, inspect, or maintain the product accordingly.

Standards data is therefore used to inform the end product—whether it be a design specification, a work instruction, or a product definition. In the past decade, the aerospace industry has taken advantage of online capability for the procurement and use of standards, and technological capability for standards as data is rapidly advancing as the digital world develops.

One rapidly emerging trend in standardization is the development and consumption of standards content. For many decades, standards were paper only: typed and printed, then provided to customers as paper copies. Later, standards data was also held on microfiche. For the drafting of standards, the incorporation of comments from the approval phase had relied on manual input and therefore a strong manual configuration management of comment resolution in a pre-digital world.

In the 1980s and 1990s the CD revolution allowed for standards products to be much more easily transported and used on networks: and CD technology also enabled standards organizations to lock the content for security and intellectual property (IP) purposes.

More recently and still today, the majority of aerospace industry standards are being used in PDF format. However, the digital age is rapidly transforming the use—and ultimately the development—of standards. Standards as digital data, including the conversion of text to XML, is enabling the aerospace industry to directly incorporate the definition and data in the documents into workflows without the need to scan and import standards data independently and from single-entity documents with the issues of integration and interoperability that this may bring. An example is the 3D parts rendering that AIA introduced during a major revision of the AIA standards portfolio: files that can be directly integrated into CAD or other systems from the standard. Dynamic redlining provides the user with a clear view on differences between revisions and the options provided by using industry-recognized data standards such as XML for tagging and metadata allows for much greater search and utilization potential for the user. The future brings yet greater promise as "standards as digital data" is fully realized to bring a seamless integration from content development to integration into a digital thread shared by multiple stakeholders.

Conclusion

The civil aerospace sector is built on standards, particularly in addressing the global and safety-critical nature of flying. And as aircraft become more varied, with modern technologies being demonstrated on dynamic vehicle types and platforms, with drones, unmanned, and autonomous technologies, with digital manufacturing capabilities, and through the pioneering spirit of commercial space, standards are becoming an ever more important facilitator of product development and the public—private partnership.

References

1. Xie, Z., Hall, J., McCarthy, I.P., Skitmore, M. et al., (2016). "Standardization Efforts: The Relationship between Knowledge Dimensions, Search Processes and Innovation Outcomes," *Technovation Innovation and Standardization* 48–49: 69–78.
2. https://www.faa.gov/regulations_policies/rulemaking/recently_published/media/Part23_FinalRule_2120-AK65_WebCopy.pdf.
3. Iwata, S., "Weights and Measures in the Indus Valley," *Encyclopaedia of the History of Science, Technology, and Medicine in Non-Western Cultures* 2nd edition Helaine, S., (Springer, New York, 2008), 2254–2255, ISBN 978-1-4020-4559-2
4. Wang, P. (April 2011), *A Brief History of Standards and Standardization Organizations: A Chinese Perspective (PDF)*, Honolulu: East-West Center Working Papers
5. http://www.sae.org/about/general/history/.
6. http://www.sae.org/about/general/history/.
7. https://www.rtca.org/content/mission-history-0.
8. Spiegelhalter, 2011.
9. http://www.eascg.eu/.
10. https://www.ansi.org/standards_activities/standards_boards_panels/amsc/.
11. https://share.ansi.org/Shared%20Documents/Standards%20Activities/UASSC/UASSC%2017-003,%20Summary%2019%20May%202017%20ANSI%20UAS%20Meeting.pdf.
12. DOD 4120.24-M, (2000), *DSP Policies & Procedures*, Office of the Undersecretary of Defense (Acquisition, Technology and Logistics), USA
13. http://www.sae.org/standardsdev/military/milperry.htm.
14. https://www.astm.org/SNEWS/NOVEMBER_2001/saunders_nov01.html.
15. https://www.nats.aero/services/challenges/civil-military/.
16. https://www.faa.gov/regulations_policies/airworthiness_directives/.
17. https://www.faa.gov/regulations_policies/rulemaking/recently_published/media/Part23_FinalRule_2120-AK65_WebCopy.pdf.

CHAPTER 12

The Future of Civil Aerospace: Onward and Upward

Ravi Rajamani
drD2 Consulting
Steve O'Flarity
Pratt & Whitney

> *I confess that in 1901, I said to my brother Orville that man would not fly for fifty years...Ever since, I have distrusted myself and avoided all predictions.*
>
> Wilbur Wright (1908)

Peering into the future always yields hazy results, with the haze thickening as the horizon recedes. In this chapter, we look at the current state and emerging trends. The civil aviation industry is on relatively solid financial footing as of this writing, with 2016 marking the seventh consecutive year of profitability [1], and there seems to be a growing appetite for change. Engineers are rethinking decades-old aircraft and engine architectures. Looking out to about 2050, the aviation sector might be substantially different from today. The future may also reflect technology catching up with old ideas that were simply ahead of their time.

Step improvements in gas turbine fuel and noise performance are reaching the market, with significant new advancements under development. Unique airframe concepts are being tested in wind tunnels and in subscale flight tests. Some are variants on the tried-and-true tube-wing architecture, and others are significantly different. New materials are being incorporated into engines and airframes for enhancements in weight, strength, and fatigue resistance. Electric propulsion is being explored. Several companies and National Aeronautics and Space Administration (NASA) are working on supersonic transport (SST) designs. Very small remotely piloted commercial drones featuring all-electric (AE) vertical takeoff and landing (VTOL) capability are readily available. That concept is evolving into larger vehicles which can deliver heavier payload over greater distances. In the United States, in the 1920s, commercial air transport got its big push from carrying cargo and mail, not people. Analogously, the growth of online retailers is creating demand for faster, lower cost methods of package delivery to nearly any place on earth, a problem addressable by autonomous VTOL cargo vehicles. Perhaps, as open-cockpit, bi-wing, piston-powered mail planes in 1927 evolved into the B747 by 1969, so might today's small remotely piloted electric drones by 2050 evolve into something completely different! Whatever this evolution looks like, we are certain of one thing: it will be an ever more exciting time for civil aviation.

For much of the 20th century tube-and-wing aircraft with podded engines mounted on the fuselage or beneath and forward of the wings were the standard fixed-wing configurations and remain so today. These designs allow lift and control surfaces to be separated from the propulsion devices, although the cylindrical fuselage creates drag without contributing to lift. What is lost in aero efficiency is made up for by the utility and pressure vessel integrity that a large cylindrical payload space provides. It is also an ergonomically satisfactory design. For example, most passengers can see out a window. However, this configuration is being reassessed, and novel structures are being envisioned to make the airplane more efficient, ideally without sacrificing advantages of prior designs.

Similarly, for the last seven decades or so, the gas turbine engine has dominated commercial aviation as the most efficient means of converting fuel to thrust at high altitudes and speeds, providing long range with large payloads.

Just as important as the technical challenges are operational and financial ones. How industries, governments, and regulatory agencies address these challenges will determine how the future of aviation will unfold.

The Current State

Aircraft and engine development are extremely lengthy processes that result in products operating for decades, and frequently require additional decades to financially break even.

Several major factors influence civil aviation development:

- Safety regulations
- Environmental regulations
- Manufacturing and purchasing economics
- Operating economics
- Future unknowns—energy prices, economic conditions, conflicts, pandemics, etc.

In most important metrics—passengers and goods transported, financial investment, environmental impact, and contribution to globalization—passenger and cargo aircraft over 40 tons (~80 seats and up) dominate market spaces. In 2015, 3.6 billion passengers flew on almost 35 million scheduled flights through 53,000 city-pair routes. By value, one-third of all global trade travels by air [2]. Airliner-class aircraft also generate 98% of all civil aviation carbon dioxide (CO_2) emissions [3].

Due to the influence factors above, barriers to entry for manufacturers at airliner sizes are daunting. Development is extremely capital-intensive and can be nation-state-scale endeavors, thereby dampening rapid or radical innovation. Airlines and cargo transport companies are the equipment customers and are necessarily risk-averse. The pace of *fielded* change is throttled by these factors.

Civil aviation is a market-driven undertaking requiring financial viability. Unlike government-related aircraft usage, commercial end users need to break even in the long run or disappear from the marketplace. Costs and revenues are key parameters, and new technologies need to buy their way in. So what is important to operators?

- Acquisition costs—purchase or lease
- Operating costs—fuel burn, crew, maintenance, financing, airport fees, etc.
- Noise costs—airport tariffs and restrictions
- Emissions costs—carbon taxes/credits
- Operating revenue—seats, range, speed, routes, passengers, and services

FIGURE 1 de Havilland Comet 1.

FIGURE 2 Airbus A319neo.

Seven decades ago, the De Havilland Comet 1 (Figure 1) was unveiled as the first pressurized commercial airliner powered by gas turbine engines. About the size of today's Airbus A319 (Figure 2) and Boeing 737-700 aircraft, it seated 36-44 passengers, required four cockpit crew members, and had four Whittle-style turbojet engines embedded in the wings. It cruised at 400 knots with a 2,400 km range. It also had major structural and fatigue shortfalls. The Comet broke new ground but only 114 were built, of which 25 were involved in accidents resulting in aircraft loss [4]. Its size similarity to today's most ubiquitous airliner families make for a reasonable comparison of technological and commercial evolution.

Several trends over the last 65 years are revealed in Table 1. (Note that single-aisle aircraft data in the table is a combination of Boeing 737 MAX 7 [5] and Airbus A319neo [6] specifications.) Today's airline economics rewards packing in as many passengers as possible. With only 20% more length and almost identical wingspan, the modern airliner seats up to 128 more passengers than the Comet, while electronics and automation cut the cockpit crew in half. The new jets cruise faster, travel almost three times as far, and do it on two fewer engines. Long range is economically advantageous to enable transoceanic routes and

TABLE 1 Progression of single-aisle airliners

	de Havilland Comet 1	Current single-aisle aircraft	Change
Entry into service	1952	2016-2017	65 years
Length (m)	28	34-35	+18-21%
Wingspan (m)	35	36	+3%
Gross takeoff weight (kg)	50,000	72-75,000	+50%
Passengers (max)	44	172	+290%
Cockpit crew	4	2	-2
Engines	4	2	-2
Cruise speed (knots)	400	450	+12%
Range (km)	2,400	7,000	+192%
Max continuous thrust (kgf)	9,000	25,000	+178%
Cruise TSFC (kg/h/kgf)	1.02	0.53	-48%

connect more destinations. Since takeoff is the flight segment of highest fuel burn, cruising further each flight increases overall efficiency. Those engines burn nearly 50% less fuel per kilogram of thrust produced. Overall industry fuel burn progress is illustrated in **Figure 3**.

The new engines are also dramatically cleaner, producing far less of the currently regulated pollutants (nitrogen oxides, carbon monoxide, unburned hydrocarbons, and smoke), as well as carbon dioxide. Without doubt, engine and aircraft reliability and safety have improved dramatically. Today, two highly reliable engines are typically sufficient to make long transits while being up to several hours away from an adequate airfield in the event of an engine or aircraft malfunction.

In addition to aircraft cockpit enhancements, advanced materials such as organic composites (e.g., carbon-reinforced polymer), aluminum lithium, titanium aluminides, and others have contributed to stronger and lighter airframes and subsystems that are

FIGURE 3 Average fuel burn [7].

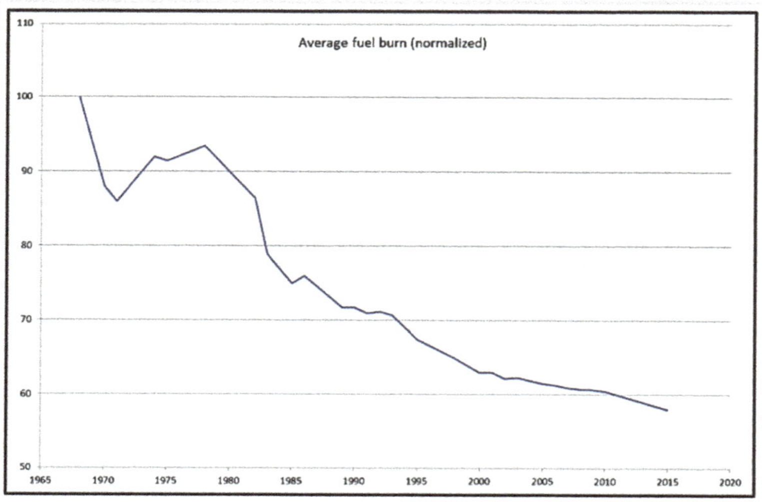

more fatigue and corrosion resistant. Advances in manufacturing technologies have led to faster and less expensive manufacturing processes with enhanced precision and reduced variability and defects.

Civil aviation includes many more aircraft operated in the general aviation (GA) sector. GA has come to mean civil aviation operations that are not scheduled air services (passenger or cargo), nor are they nonscheduled air transport operations for remuneration or hire (such as a charter service). These are typically small corporate or personal aircraft used for business or pleasure. According to the General Aviation Manufacturers Association (GAMA), the United States has around 19,000 GA airports (public and private), while scheduled services operate from about 500 airports [8]. Aircraft may be single or twin propeller-driven or jet propelled. Piston or gas turbine engines may power the propeller aircraft. The United States has more than 220,000 GA aircraft (from amateur-built aircraft, balloons, and rotorcraft to complex jets) [9]. Smaller aircraft tend to have single piston-powered engines driving open propellers and operating on aviation gasoline, which contains significant quantities of lead. Higher-end (in performance and price) business aircraft tend to be twin or triple jets with turbofan engines. They may cruise at altitudes of 40,000 to 50,000 ft, at speeds up to Mach 0.9, and ranges of 7,000 nautical miles or longer. Unfortunately, on average in the United States, GA accidents claim one life every day. Relative pilot inexperience in challenging situations is a leading cause. Regulatory bodies have ongoing initiatives to improve GA safety.

One forecast suggests that from 2017 to 2036, annual passenger traffic growth will average 4.7%. This will require growing the current passenger jet fleet from 23,480 aircraft to about 47,000, requiring significantly expanded airport infrastructure worldwide [10]. The single most significant barrier to the growth of the airport infrastructure is noise, but great progress has been made in alleviating noise, particularly with the recent generation of single-aisle engines. The new Pratt & Whitney (P&W) Geared Turbofan™ (GTF™) engine incorporates a reduction gear that enables the fan and the rest of the low- pressure rotor to rotate at different speeds. This results in a 70% reduction in noise footprint on takeoff, as well as a 16% reduction in fuel burn. The fan diameter is larger, bypassing more air around the core for greater propulsive efficiency, while turning much slower for reduced noise. One result is that, at flyover conditions, engine noise is lower than airframe noise. **Figure 4** illustrates this progress. The A320ceo (red) utilizes the previous generation of turbofan engines. The A320neo (turquoise) utilizes new-generation engines. This latest generation of engines provides a remarkable reduction in community noise impacts. There is, however, more to do.

FIGURE 4 Comparative takeoff noise footprints for five aircraft.

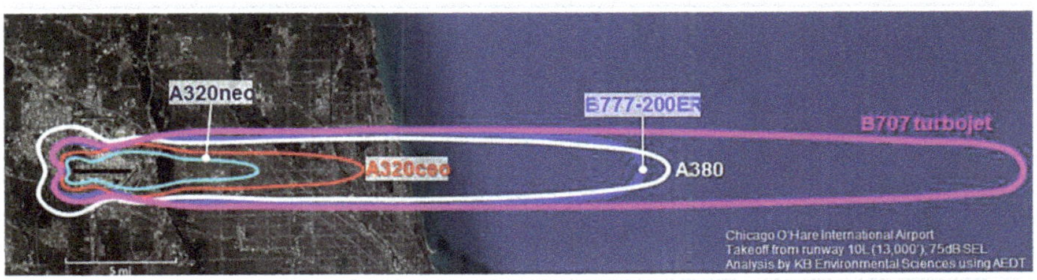

Propulsion

Those who might believe gas turbine performance has peaked would be mistaken, as illustrated in Figure 5. It shows overall thermodynamic efficiency (converting fuel energy into rotational power) and propulsive efficiency (converting rotational power into moving air—thrust). The product is overall efficiency which is directly proportional to thrust-specific fuel consumption (TSFC) or fuel efficiency. The points represent performance of engines that were or are in commercial service. Frank Whittle's engine demonstrated an overall efficiency of about 10%, well off the top of the chart. The latest production gas turbine engines improve upon that by a factor of four. Yet, most engine manufacturer experts believe another 20 points improvement in overall efficiency is possible, resulting in a further 50% fuel efficiency improvement from today's best.

How will this be achieved? One way is reducing cooling air through more effective cooling strategies, cooling the cooling air, or eliminating it altogether with materials that don't require it. Such materials could include composite or monolithic ceramics. As thermal efficiency rises, core size for a given thrust class shrinks. To boost propulsive efficiency, bypass ratio (BPR), the amount of fan air bypassing the core versus the amount used by the core, will continue to increase. These characteristics are evident in the geared turbofan engine. The fan diameter is very large and the core very small. The P&W GTF engine has the highest propulsive efficiency of any single-aisle aircraft engine, resulting in a step improvement of 15%–16% in fuel burn relative to comparable models before it. Harry Truman is reported to have said "There is nothing new in the world except the history you do not know." As if to prove Harry Truman correct, the GTF is not the first

FIGURE 5 Historical gas turbine efficiency vs. achievable capability.

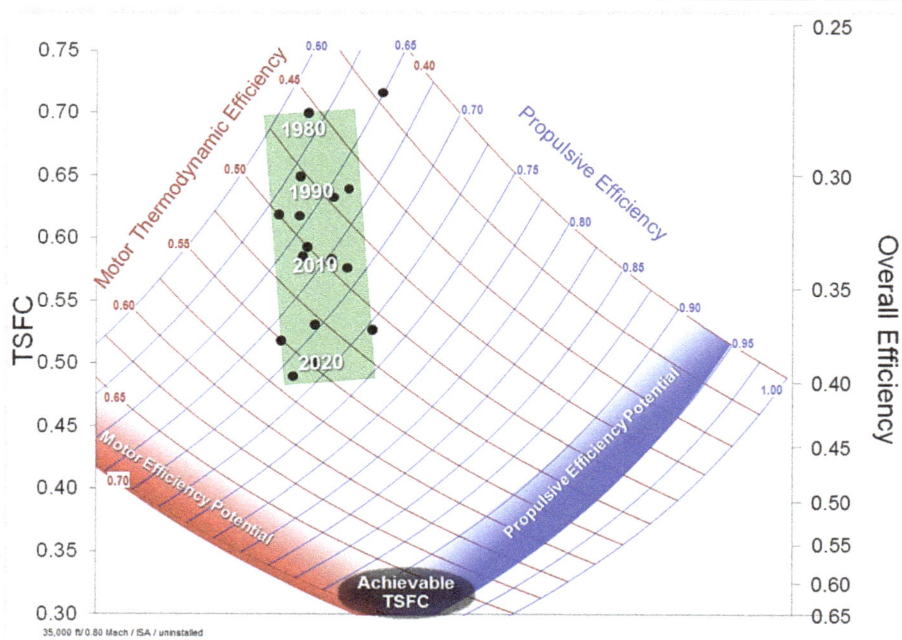

engine to use a geared arrangement; many developmental attempts were made. The first production geared engine was the Garrett TFE731 in 1970. The P&W GTF engine is the first large engine to make a reduction gear system efficient (99.5%) and reliable. Not only does it enable a slower-turning, more efficient fan, it has a similar effect on the low-pressure turbine (LPT). In general, turbine efficiency is enhanced at high speeds, and the GTF LPT is more efficient while requiring fewer stages, resulting in less weight and lower manufacturing and maintenance costs. It is likely that most or all new airliner turbofans will be of the geared architecture for the foreseeable future.

The first recognized automobile used a single-speed belt system between the motor and drive wheels. In a traditional turbofan analogy, the core stream constitutes the motor, the fan, and the driven wheels. It's a fixed, direct drive system. The GTF is the same except it breaks the low-pressure rotor in two, with speeds determined by a fixed gear ratio. When a clutch was introduced in the automobile, it enabled the operator to disconnect the engine from the drive wheels for beneficial reasons. Aviation now has that in the F-35B aircraft with a clutch capable of handling about 25,000 hp to engage or disengage its hover "wheel," the vertical lift fan. Perhaps these breakthroughs give us clues about the future. Today, autos have electronically controlled automatic gear systems with many selectable ratios to optimize operation under different conditions and "missions." Drag racing cars have compact variable-ratio gearboxes capable of transmitting 10,000 or more horsepower. Mightn't a future gas turbine engine also incorporate some form of variable-ratio gear system into its drivetrain? This could allow in-flight tuning of fan speed independent of core and LPT power for variable bypass ratios.

The fan in a turbofan engine is not much more than a propeller inside an open-ended cylinder. Modern propellers typically have propulsive efficiency of 85%–90%. The best gas turbine fans are in the low 70s, so there's room to improve. The fan case serves two principal purposes—noise reduction and protection against fan blade failure. Elimination of the fan case is a technology challenge in large commercial aircraft, but in so doing, fans may morph into "open-rotor" configurations, usually counter-rotating swept propellers, with higher propulsive efficiency (Figure 6). When or if commercially viable open-rotor configurations come to market, this will again be nothing new. Commercial open-rotor engines were flight-tested in the late 1980s, and military versions, wherein noise may be more tolerable, have been in production since at least the 1950s. Like the GTF engine, a successful open-rotor gas turbine engine will be the result of technology catching up to an old idea.

FIGURE 6 Counter-rotating open-rotor test, 2017.

Smaller gas turbine engines, such as those up to about 10,000 lb of thrust, and all turboprop engines, typically have lower component and overall efficiencies than airliner-class turbofans. The reasons relate to the cost of technology and the market space. Smaller engines require smaller parts with smaller features, such as turbine airfoil cooling holes and operating clearances in compressors, which can lead to significant manufacturing, and hence cost challenges. Many small engines are designed for very price-sensitive low-utilization markets, making it economically unappealing to put exotic materials or manufacturing techniques into these applications. Because the total volume of a cylinder reduces as a cube of the corresponding linear measure, while the frontal area only reduces as the square, multiple small engines may provide the same power as one large engine while requiring less total volume. This can lead to architectural flexibility and may also lead to cost economies of scale, simpler maintenance, and other benefits. Given that the trend toward ever greater BPR, and thus larger fans, will inevitably lead to more challenging or prohibitive installations, we may see a reversal of the shrinking-engine-count trend, toward multiple commodity small engines that are cheaper and easier to integrate. There is no physics-based reason why small engines cannot be as efficient as large ones. Multiple engines can also be used to assist flight controls thereby possibly removing control surfaces altogether in a future aircraft architecture. Affordable manufacturing technology just needs to catch up.

Gas turbines are so effective because of the extremely high-energy density and easy handling of liquid jet fuel. Commercial engines worldwide operate on kerosene-based Jet A fuel or its close cousin Jet A1, which are largely refined from petroleum. Global infrastructure and airline economics make it nearly impossible to introduce a new fuel that is not Jet A fungible. In 2017, global commercial airlines consumed 88 billion gallons of fuel [11]. And while gas turbines can theoretically operate on almost any liquid hydrocarbon, such is not achievable in practice. Specifications for jet fuel include many requirements to ensure safe operation of engines, aircraft, and infrastructure under nearly any environmental condition or circumstance.

Every mass unit of jet fuel combusted in air releases about 3.15 units of CO_2, a greenhouse gas linked to global climate change. Increasing pressures, societal and regulatory, to reduce aviation's carbon footprint have led to development of sustainable alternative jet fuels (SAJFs), many of which are plant-based. Today, the approved fuels that meet the same specifications as Jet A are blends of a particular SAJF and Jet A. They are drop-in fuels, completely fungible with Jet A. Quantities are low—the 2017 delivery estimate is about one million gallons [12]—and costs are high, but this will likely change, and the future will see a large production capacity of pure SAJFs that do not require blending with petroleum-based fuel. On a life-cycle basis, bio-derived SAJFs may reduce civil aviation's GHG contribution by 80% or more. That, combined with engine and aircraft improvements and operational enhancements, such as elimination of step descents in favor of continuous descents, will provide for carbon-neutral growth and ultimately steady reductions.

Fuel also plays a part in the future of the 230,000 aircraft worldwide using piston-based propulsion. Today, the primary fuel is aviation gas (avgas), which incorporates lead to reduce damaging preignition/detonation of the fuel-air charge. In the United States, avgas is the largest contributor to lead emissions. When ingested in any form, lead is very harmful to human health [13]. Were lead to be banned, the much lower octane level would lead to engine failures. Research is ongoing to find a replacement fuel that provides the same functional characteristics without lead. It's a major challenge because avgas is so well-matched to piston engine altitude-performance requirements. Eventually the future of GA will include the phaseout of lead-containing fuel, ideally with a drop-in bio-based replacement. It's possible the longer-term future will witness the phaseout of piston engines altogether.

Another resurrected trend is the desire for speed. Business and nonbusiness travelers alike generally prefer to spend less time in the air. While many military supersonic aircraft have been developed, the one major civil example, Concorde, was a commercial failure. If the barriers to entry for new subsonic commercial aircraft entrants are high, they pale in comparison to those required to develop an economically viable supersonic transport, even when supported by national governments as the Concorde was (United Kingdom and France). Development began in 1960 and it entered service in 1976. Only 14 production aircraft were built and flown by the two countries' national airlines. Development and production costs were many times original estimates, four thirsty turbojet engines burned substantially more fuel per passenger-km than other aircraft, and ticket prices were generally only affordable by the rich. Numerous noise restrictions limited available routes. However, any aircraft is a compromise of conflicting requirements, and the Concorde design teams overcame many obstacles to get the aircraft certified and in service at all. It came to an end with the last flight in 2003. Many sources of information on the successes and failures of the Concorde program can be found. One broad takeaway is obvious: even with governmental support and some of the world's most capable engineering talent, it is still incredibly difficult to develop a passenger aircraft that meets all the requirements of mature subsonic aircraft, is sufficiently environmentally friendly, and is a commercial success, while cruising 50%–100% faster. Fortunately that doesn't stop people from trying.

Several private entities are still working on designs for supersonic passenger aircraft with seat capacity from large business jets to small airliners, less than the Concorde's 92–120 passengers. This size may appeal to a broader customer base. NASA is funding a Lockheed Martin demonstrator project, and flight testing and wind tunnel work to address sonic boom intensity has been undertaken

There is a saying that the best aircraft can only be built around the best engine. This is especially true for aircraft intended to cruise at supersonic speeds while being economical and meeting noise regulations. The propulsion challenge is conceptually simple. A propulsion system is needed that can efficiently operate in two distinct regimes. The first regime operates with high airflow and high overall pressure ratio (OPR) at low speed and altitude—takeoff, approach, and landing. This minimizes noise and maximizes takeoff power without the use of an afterburner, operating much like a conventional airliner engine with a big fan and small core. Takeoff power is especially critical because the supersonic airframe has many challenges, at least some of which lead to relatively small wing area, resulting in higher rotation airspeed than conventional airliners. In the second regime, the engine needs to operate at high velocity and altitude more like a low BPR turbofan with low OPR and a small inlet. Low or no bypass (i.e., turbojet engines) minimizes ram drag and maximizes specific thrust (thrust per unit of inlet airflow). At supersonic speeds, an air-breathing engine performs like it has additional compression stages due to the ram effect, heating and pressurizing incoming air. Low OPR, while less efficient, is required to manage internal temperatures and pressures. Low BPR/low OPR favors a small (or no) fan and large core. This is the realm of military fighter engines.

Since an ideal engine doesn't yet exist, some prior concepts utilized military-derivative engines, as did the Concorde, sacrificing noise and low-altitude fuel burn to meet the supersonic cruise objective. Other designs considered older low-bypass commercial engines like the P&W JT8D, which would be challenged by the continuous high-temperature cruise duty cycle and noise. More recently, midsized commercial engine derivatives have been proposed using reduced fan size and fewer low-pressure compressor (LPC) stages to decrease BPR and OPR, still a compromise between takeoff thrust, noise, and supersonic cruise capability. In developing proof-of-concept prototypes and/or production vehicles, the three-engine manufacturers capable of designing and developing suitable engines are reluctant to do so due to very high costs, air vehicle risk, and a small potential market. Regardless, one contender, Aerion Supersonic, is developing a three-engine Mach 1.4 supersonic business jet, the AS2 (**Figure 7**). The engines are planned to be based on a General Electric (GE) Aviation CFM56 engine core, adding a new low rotor with a smaller diameter two-stage fan for lower BPR/higher fan pressure ratio (FPR). To counter the takeoff noise issue, part-power takeoff is proposed, which will also lengthen takeoff distances [14].

Is a variable-cycle civil engine conceivable? Yes, but it must be light, compact, low drag, and affordable. It may involve maturation and implementation of a collection of technologies to vary OPR and FPR. This could include fan bypass, combustor exit area, exhaust nozzle area, cooling flows, etc. In addition, variable inlets may be used to control shock waves and inlet flows at varying flight speeds.

The only operational variable-cycle engine was the late 1950s era (**Figure 8**) for the Central Intelligence Agency (CIA) A-12 and U.S. Air Force (USAF) SR-71 reconnaissance aircraft, which could transition from a turbojet to a ramjet, enabling Mach 3.2 cruise speeds above 90,000 feet altitude. The J58 was a single-spool engine with an eight-stage axial compressor can-type combustor, two-stage axial turbine, afterburner, and variable exhaust nozzle, typical of military fighter engines of the day. Airflow was high and OPR very low. What was not typical was the series of large valve-regulated tubes that could bypass airflow from the middle of the compressor directly into the afterburner. At takeoff and climb, it functioned just like a typical turbojet with afterburner off or on. At high speeds and altitudes, the bypass tubes would open, reducing flow around the back of the compressor, combustor, and turbine into the afterburner/nozzle section, thus becoming a ramjet. The core was doing almost no work, OPR was reduced by two-thirds, with most of the airflow and pressure rise being provided by ram compression into the inlet to be burned with fuel in the exhaust, like a ramjet. At cruise, temperature at the engine face was 800°F. No engine before or since has been capable of such performance.

In many ways, the J58 engine was phenomenally advanced for its time, particularly considering it was developed in the age of slide rules. Going beyond 1.5–2.0 Mach in a civil SST could be enabled by modernizing the J58 architecture. Materials, coatings, aerodynamics, combustors, controls, and so on have

FIGURE 7 Aerion AS2 aircraft.

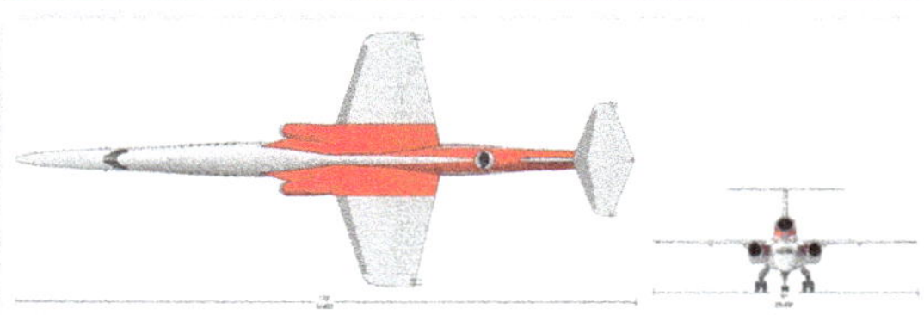

FIGURE 8 Pratt & Whitney J58 variable-cycle engine.

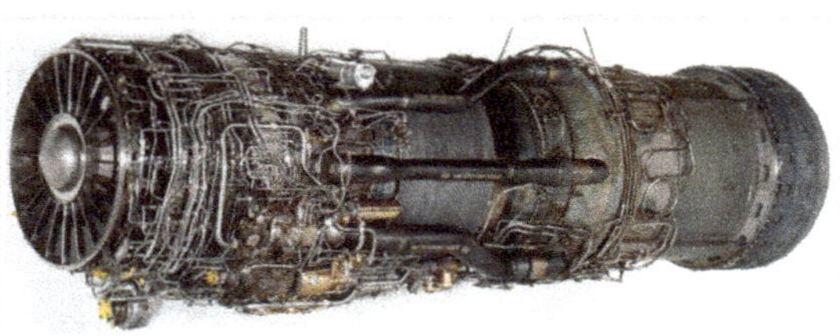

been greatly improved. Using today's and tomorrow's tools and technologies, it is very likely that a "J58 2.0" could be developed that is lighter, more compact, and more efficient. Commercial Mach 4 cruise could be within reach.

Sub- or supersonic aircraft might utilize distributed propulsion—dividing up a large unwieldly fan into multiple small ones—driven either mechanically or electrically. Figure 9 depicts one of the earliest efforts toward distributed propulsion, the 1940 modified Allison V-3420-B-5 engine driving two propeller shafts. Figure 10 shows a 2017 concept with one gas generator mechanically driving four fans. As fans get larger in search of better propulsive efficiency, they will become harder to install under wings, heavier, and with higher drag. The configuration of Figure 10 would be difficult to install on a conventional cylindrical tube-and-wing airframe. Other configurations might be more amenable, to be discussed in section Airframe.

FIGURE 9 Distributed propulsion, 1940 [12].

FIGURE 10 Distributed propulsion, 2017.

Boundary layer ingestion (BLI) is another engine-airframe integration tool available to potentially improve efficiency. In a BLI application, the airframe is configured to allow fans or engines to ingest some of the high-drag fuselage boundary layer and reenergize the wake with engine exhaust, reducing overall drag. Implementing BLI requires a highly integrated propulsion system with fans that can swallow the high levels of turbulent flow without structural or aerodynamic shortfalls. Figure 11 depicts a 2017 test of a highly distortion-tolerant fan in a BLI configuration in a NASA wind tunnel.

FIGURE 11 2017 BLI/distortion-tolerant fan test.

In recent years, interest in aviation electric propulsion has substantially increased. This is certainly inspired by advancements in and acceptance of practical and cost-effective hybrid-electric (HE) and AE models, many of which have been rolled out. Once again, good ideas frequently wait for technology to implement them. The first electric-powered vehicle was built by Robert Anderson in the mid-1830s [17].

The need for manufacturers to achieve corporate average fuel economy (CAFE) requirements and the Tesla charisma effect are also contributors to the expansion of electrically motorized cars. While HE vehicles (with an internal combustion "range extender" engine) have been available since the late 1990s, pure plug-in automobiles have only been successfully mass-marketed since the mid-2000s. Tesla S, the most notable example, went on sale in 2012.

Electrically propelled cars have substantial advantages over aircraft. Cars roll on the ground without need to overcome gravitational force. Automotive travel involves slowing and stopping, things that normally do not happen in flight. An electric car can recharge batteries by capturing energy dissipated through braking, regeneratively converting it to electricity. Lastly, since cars cannot fall from the sky, and getting out typically only requires stopping, then ensuring adequate drivetrain reliability and safety is a much simpler task than for civil aircraft. Automotive recharging is potentially available any place with an electric socket. "Range anxiety" takes on a more urgent meaning in an airplane.

Why would we want to switch to some type of electric propulsion? One frequently cited answer is to reduce aviation's atmospheric CO_2 contribution. However, an HE aircraft propulsion system would still use a thermal engine for propulsion or battery charging, which still produces CO_2. An AE system requires charging batteries, typically from the electric grid, which is still powered principally by fossil fuel. However, with the growth of renewable sources of power, the overall environmental benefits of AE systems are growing. Worldwide, two-thirds of electricity originates from fossil fuel, most of which is less thermally efficient than large turbofan engines. An electric system can never be cleaner than its source of electricity. However, most proposed HE and AE aircraft are in the GA category, and as previously noted, to move the needle on aviation CO_2 production requires addressing airliner-sized aircraft, not regional or business. A comprehensive account of the history and the technologies involved in electric propulsion is available in a recently published book [18].

Another proposed benefit is noise reduction. It's true that electric motors can be quieter than thermal motors. However, for aircraft we also need a propulsor to convert rotational energy into thrust to fly. For smaller planes, an electric motor would be connected to a propeller, a substantial noisemaker. For a larger aircraft, it might be driving a conventional fan. In either case, the predominant noise source is the propulsor, regardless of the motor type. In some applications today, airframe noise exceeds total engine noise in flyover and landing. Noise reduction is not a compelling justification for electric propulsion motors except in piston engine and small gas turbine applications.

Some believe changing to electric will reduce operating costs, predominantly fuel or maintenance. Due to engine inefficiencies and small fuel purchasing volumes, small aircraft operators can have high fuel bills, and a potential electric benefit is plausible, although it can also be negated by additional weight of the electric propulsion system. Aircraft electric propulsion system maintenance costs are yet unknown.

FIGURE 12 Pipistrel Alpha Electro trainer.

Safety is the principal requirement for an aircraft to be certified by regulators. Beyond certification requirements, aircraft capability—payload and range—is largely defined by power and energy on board per unit weight. Weight is the primary "enemy" in aircraft design and manufacturing. Most flight-capable birds have hollow bones for lightweight. As of this writing, several small aircraft have been converted to operate on battery-electric propulsion systems. They typically carry two people and have endurance measured in tens of minutes. Consequently, the initial uses are as motorized gliders and flight training in and out of a fixed base, e.g., Pipistrel Alpha Electro (**Figure 12**), which is certified as a light aircraft trainer in several countries. Small aircraft may be the entry point of electric propulsion into civil aviation for two reasons—flight-weight systems only exist at the smallest aircraft sizes, and piston and small gas turbine engines are less than half as efficient as large turbofan engines, so an energy cost benefit is more credible.

Flight-weight electric propulsion motors have been developed up to about 250 kW power level with a gravimetric power density of about 5 kW/kg. Practical battery packs (including controls, safety, charging management, and environmental protection requirements) have a gravimetric energy density of about 0.150 kWh/kg [19].

When thinking about battery-electric propulsion for airliners, consider that a 150-passenger single-aisle airliner today utilizes fuel with 12 kWh/kg energy content, 80 times that of battery packs. It requires 36,000 kW to take off, with gas turbine motors (everything but the fan, equivalent to the electric motor) producing about 12 kW/kg of power. To fly a 1,000 nautical mile mission, that airliner will require about 50,000 kWh of energy. This represents about 4,200 kg of Jet A, or about 333,300 kg of today's battery packs. Even batteries 10 times as energy-dense would weigh 33,000 kg, a performance level forecasted to be decades away. If there was a technological breakthrough leading to viable battery-powered airliners today, the infrastructure cost to replace 88 billion gallons of Jet A usage per year with electricity generating capacity would be on the order of half a trillion dollars.

Superconductivity, utilizing a cryogenic fluid to reduce electrical resistance, has potential to reduce weight of high-voltage cables, motors, and generators. Cost and complexity of such a system are daunting, and it is known that failure of the cryogenic cooling in an operating system can result in catastrophic consequences. It does have potential that is not likely to be realized in civil aviation propulsion before 2050.

Urban mobility and cargo drones are potential applications of HE and battery-electric technology, more suitable in the near term due to size. Some small cargo delivery applications, similar to existing commercially available quadcopter-like vehicles, are in advanced testing stages. Larger urban mobility solutions are technologically feasible. Broad implementation will require substantial regulatory and air traffic management work, as well as logistical considerations—access locations, charging/refueling, and so on.

In our view, the electric propulsion timeline is driven by application size: payload drones may come sooner, airliners much later, if ever. Scaling up electric propulsion components to large sizes while simultaneously increasing energy and power density, lowering costs, and building out the required infrastructure is likely to be beyond the time horizon contemplated herein.

Airframe

As one could infer from the propulsion discussion above, engine-airframe integration is becoming ever more critical to enabling advancements. Supersonic flight is a case in point. A conventional tube, wing, and podded-engine aircraft will have prohibitively high drag to breach the transonic region in level flight. To transition through the Mach and cruise supersonically requires very low drag. Hence, modern civil SST designs have very long, narrow fuselages and relatively small thin wings with greater leading-edge sweep than conventional passenger aircraft. Given that architecture, seat number is highly constrained. Given an engine as described in section Propulsion, then the challenges come down to these. Designing a super-cruising aircraft is surmountable. The principal technology barrier is overcoming noise—takeoff and sonic booms. Takeoff noise can be managed through operational and configurational techniques such as engine shielding, adding sound-deadening material, slower takeoffs, and so on. Supersonically, much work has been done on shaping the aircraft to shrink the boom overpressure such that it reaches the ground at very low levels or not at all.

Business challenges are more problematic. It is essential that key routes be allowed, which may require multiple regulatory changes in several countries. Aircraft manufacturing and acquisition costs must be kept acceptably low. A supersonic aircraft may have substantial challenge meeting current or future emissions regulations. And it will have a limited passenger capacity due to the narrow fuselage (Figure 13), limiting revenue in a civil application. The larger market may be the very high-end private aircraft market. A 2018 census listed over 2,200 billionaires worldwide, many of whom would gladly purchase such a rapid mode of global travel and ultimate status symbol. Summarizing, the hurdles are noise, regulation, and financial viability.

Rotary-winged aircraft, typically helicopters, have been undergoing technological progression through demonstrator concepts initially aimed at military applications but well-suited for civil use. With single-rotor traditional helicopters in stationary hover, the entire rotor produces lift. The tail rotor is required to oppose the torque provided to the main rotor and keep the helicopter from spinning. To fly forward, the blades are tilted to add a horizontal thrust vector. The faster the machine flies horizontally, the less lift is produced by the retreating portion of the rotating blades, which requires helicopters to have special cyclic control to equalize the lift. But this complex control system limits speed and efficiency.

A better idea is found in application of the Advancing Blade Concept (ABC), wherein fixed-pitch contra-rotating coaxial rotors are utilized to reduce or eliminate the retreating blade stall issue in parts of the envelope. ABC also eliminates the need for the tangential tail rotor with its drag, complexity, and ground hazard. The concept was demonstrated by Sikorsky in flight testing of the XH-59A in 1975. The aircraft was quiet, stable, and fast, reaching 156 knots in level flight. By 1978, two auxiliary propulsion turbojet engines had been added, making it a compound helicopter. This had the added benefit of unloading propulsion requirements from the main rotors and boosted speed to 263 knots [3-2].

FIGURE 13 Boom supersonic jet.

As a demonstrator aircraft, the XH-59A was successful in proving the ABC rotor and compound configuration. While effective, the small turbojets were noisy and inefficient. The next step, in 2008, was the Sikorsky X2 compound rotorcraft, utilizing composite rotors and an integrated pusher propeller at the rear for horizontal propulsion. It was a much more integrated design and achieved 250 knots in level flight. The X2 led to the S-97 Raider Light Tactical Helicopter prototype in 2014, and the subsequent Sikorsky/Boeing SB>1 Defiant submittal for the Army's Future Vertical Lift program.

In 2013, Eurocopter developed a compound helicopter derived from their Dauphin aircraft. Stub wings with variable-pitch propellers were added and the tail rotor eliminated. The wings provided lift and the propellers thrust, allowing the main rotor to be slowed, reducing drag and retreating stall propensity. This unique design achieved 255 knots in level flight.

While the X2 (**Figure 14**) and X3 (**Figure 15**) were prototypes for potential military contracts, their architectures are applicable to civil markets. Noise and fuel consumption would be critical, and perhaps the X2 configuration would have an advantage there. Airbus Helicopters is developing a civil X3-architecture compound helicopter under the European Union's (EU) Clean Sky 2 program called RACER—Rapid And Cost-Effective Rotorcraft. Many smaller electric aircraft, for example, the Kitty Hawk Cora, use the same principle for faster forward motion. These and similar designs should be part of the coming generations of vertical transport.

FIGURE 14 Sikorsky X2.

FIGURE 15 Eurocopter X3.

A number of concepts for airliners departing from the ubiquitous cylindrical fuselage and wing with podded engines have been studied and some subscale models flown. These include the "Double-Bubble" fuselage with embedded aft-mounted engines in a BLI configuration. The wider fuselage provides lift, reduces trim drag, and allows room in the aft end for engines and inlets open to the top to ingest the boundary layer air (**Figure 16**). The fuselage would require internal structure to allow the non-cylindrical cabin to handle pressure loads.

FIGURE 16 Aurora Double-Bubble/BLI concept.

Aft boundary layer concept: Conventional fuselage, wings, and two engines, with a third engine and large fan at the aft end. This fan would be positioned to ingest the majority of the boundary layer around the fuselage circumference (**Figure 17**).

Flying wings: This is another invention true to the Harry Truman quote that needed more time and technology to become viable. True flying wings are tailless and have no discernable "fuselage." Gliders, propeller-driven, and jet-powered tailless swept wing designs go back to the early 1900s through World War II. Since 1940, Northrop Corporation has been at the forefront beginning with the N-1M, followed by the N-9M (first flight 1942), a one-third scale prototype of the XB-35 bomber (1946), the YB-49 jet bomber (1947), and ultimately the B-2 Spirit bomber (1989) (**Figure 18**). Flying wings are inherently more efficient because the entire aircraft provides lift and they maximize useable volume. Short length and no tail provide a stability challenge, overcome by modern flight control systems. Access (easiest through the bottom), size, and shape are hindrances to existing passenger gate access. Perhaps most significantly, side-viewing passenger compartment windows would be infeasible.

FIGURE 17 Aft BLI concept.

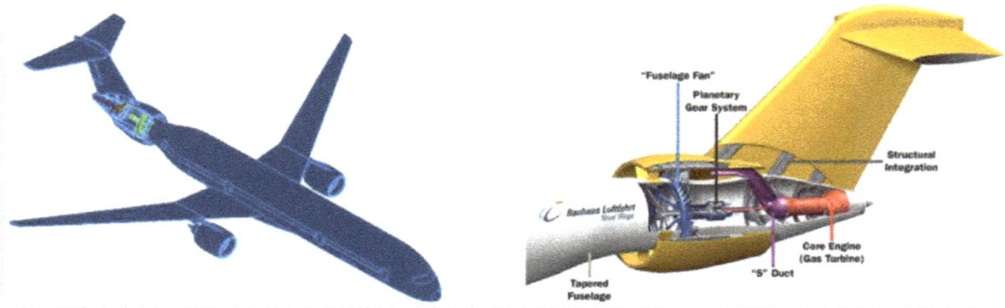

FIGURE 18 B-2 flying wing.

FIGURE 19 NASA X-48 BWB.

FIGURE 20 NASA HWB concept.

Flying wing variants: Blended wing body (BWB, **Figure 19**) and hybrid wing body (HWB, **Figure 20**). Similar to the flying wing but with aft tails and vestiges of a longitudinal fuselage, they enable a more traditional cabin while retaining some of the flying wing benefits. Flying wings and their variants lend themselves to aft-mounted engines, either podded above the aft deck or embedded in a BLI configuration. The aircraft width also supports distributed propulsion through multiple small engines (YB-49 had eight) or distributed fans. No technological roadblock exists to the development of passenger or cargo flying wing-type aircraft.

Other types intended to increase wing performance: These include boxed wings, truss-braced wings, and other similar concepts.

The principal barriers to any of these potential airliner configurations are financial uncertainty and risk avoidance. Initially, development and certification will take longer and cost more than learned-out designs, especially since some certification requirements are tailored to current designs. The cost of putting a new aircraft into production can exceed US$10 billion today. Airframe OEMs will be concerned about uncertainty of airline and passenger acceptance, and airlines and lessors are possibly even more risk-averse. Sold backlogs of existing types are out into the mid- to late 2020s, and with annual orders currently exceeding deliveries, those backlogs will continue to grow. There is very little to suggest that the industry will produce any significantly new airplane configuration before 2050 as there is little incentive to do so.

Airframers are slowly replacing hydraulic actuation systems with electrohydrostatic actuators (EHAs) and electromechanical actuators (EMAs). The Boeing B787, for example, has increased its content of electrically driven components. Neither the engines nor the auxiliary power unit (APU) provides any bleed air for aircraft functions. Functions such as engine start and environmental control have been replaced by electrically actuated systems. Aircraft will be able to take full advantage of this move to ME systems when they eliminate the hydraulic system entirely, thereby saving weight and cost associated with hydraulic pumps, valves, and so on. This will happen only when EMA components achieve the reliability of the hydraulic components they will replace. Because of their perceived lack of this high level of reliability, currently EMAs are only used to actuate non-safety-critical systems, unless multiple redundancies can be designed in, such as with brakes. This is not an insurmountable task, and the day when hydraulics can be eliminated from the aircraft is not far away.

Regional and business applications for turboprop-powered aircraft have little in the way of new airframe concepts on the horizon. The very large regional turboprop market is dominated by ATR and Bombardier with products seating 48 to 82, and the segment is shrinking. It is unlikely any significantly new airframe configurations will be introduced in this segment, unless a breakthrough in electric propulsion occurs that requires or enables new beneficial designs.

The exception might be a regional aircraft designed for HE or AE propulsion with electrically powered ducted variable-pitch fans as has been proposed by some (Figure 20). The advantage of such a system relative to wing-mounted motors driving propellers is reduced noise. Even when optimized with fixed pitch, ducted fans have lower propulsive efficiency than open propellers, and the ducts add weight. The benefits of variable-pitch fans are similar to variable-pitch props—off-design performance and integrated reverse capability. Variable-pitch gas turbine fans have not been successful in the past due to trade-offs between blade shape, duct tape, and tip clearance. To maintain constant (tight) tip clearance and flow-path continuity, the fan blades need to be perfectly straight and the inner and outer flow paths need to be perfectly cylindrical, a situation that is never found in contemporary fans due to substantial efficiency debits. But perhaps the ducted variable-pitch fan is another idea awaiting technology to get on board.

The segment of regional jets and small mainline aircraft is being refreshed at this writing, amid new entrants including Bombardier CS100 and CS300, Mitsubishi MRJ 70 and 90, and Embraer E175-E2, E190-E2, and E195-E2. These extremely modern aircraft cover the seat range from 70 to 160. All feature many of the technologies discussed herein including advanced lightweight materials and geared turbofan propulsion. They will serve the segment well for the next several decades.

The light aircraft space has long seen a variety of novel airframe configurations because of the flexibility and affordability a small aircraft can provide. Variants for all types of roles and uses abound from the ubiquitous Cessna 172 to the De Havilland Canada Twin Otter to amphibious aircraft with or without auxiliary floats. There are tricycle landing gear, retractable or not, and tail wheel landing gear. Acrobatic planes and crop dusters. We expect this boundless variety to continue while occasionally creating innovations that make their way into larger aircraft.

Urban mobility is a new airframe space that is pure innovation. Substantial capital is being invested in a multitude of seasoned and start-up companies; optimists and believers abound. Proposed vehicles are similar in concept to quad-copter recreational photo drones but designed to carry people. Many have more than four propellers, some have coaxial rotors, and others have pusher propellers for enhanced horizontal flight. Many are battery-electric, with correspondingly short flight durations, and some are HE with an additional thermal range-extending engine. Some are planned to be autonomous, while others are piloted.

For most, the target business model is air-taxi service. In all cases, the challenges are regulatory and technological. The regulatory issues likely exceed those for larger electric aircraft in that no vehicle safety standard yet exists. These vehicles will be flying in currently uncontrolled airspace much closer to the ground amid obstacles, like buildings and power lines, in urban and suburban environments. Assuming a dominant concept is rooftop passenger pickup and drop off from designated locations in a city, AE vehicles will need to recharge or swap batteries after each, or just a few, flights. HE vehicles will need to fill up with liquid fuel at least daily if not more often. Carrying excess fuel to avoid fill-ups will impact range or payload. Noisy drones will add to urban cacophony. There are places where this may be workable and places where it will require much more technology and infrastructure to have a chance. Despite decades of automotive safety and reliability development, all types of accidents and failures occur daily. Air-taxi vehicles will need to be designed to either completely avoid the equivalent of an automotive "fender bender" or be able to survive it.

Air traffic management will be problematic in a space without reasonable radar accessibility. Today, collisions between helicopters in urban areas are not uncommon, especially with helicopters covering developing news. One could envision an air traffic management system necessarily made up of all the urban air vehicles in the system equipped to be nodes in a mesh network with sensors and communication systems, perhaps augmented with sensors on buildings, to autonomously avoid all obstacles and optimize flight efficiency. Of course, this system would need to be completely hardened against cyber attacks or other malicious interference. Beyond cyber security is physical security. Airliners and road vehicles have been used as weapons. In major cities there will be very significant governmental concerns and regulations for flying vehicles operated by taxi companies.

FIGURE 21 Kitty Hawk Cora AE Air Taxi.

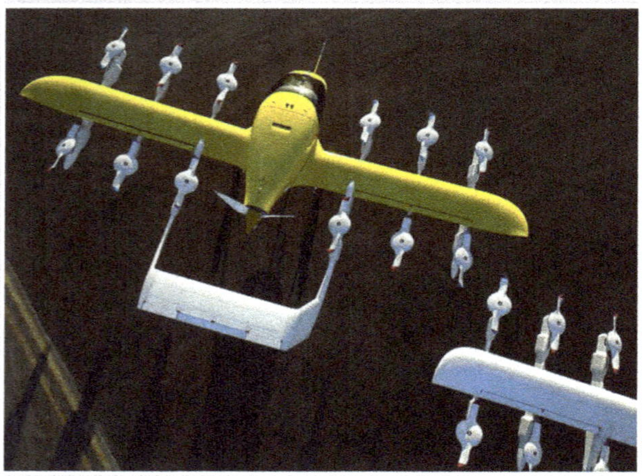

It will be challenging but over time there is a concrete chance that these challenges will be met, and air taxis will become a reality. But our crystal ball is too clouded over with all the diverse hurdles to say when it will come with any semblance of certainty. We guess that before 2050 there will be several localized or regional systems operating somewhere in the world. Some likely contenders are depicted in the accompanying figures, three of which have actually flown (**Figures 21-23**).

A concept HE regional jet design from Zunum (**Figure 24**) has not yet been built.

In the important larger transport aircraft category, AE designs are impractical with current technology and the research is concentrated on HE solutions. Airbus has teamed up with Siemens and Rolls-Royce on a test demonstrator program called E-Fan X [27]. This program aims to demonstrate the concept of HE propulsion by replacing one of the four Honeywell turbofan engines on a BAe 146 aircraft with an electrically driven fan powered by a generator attached to a smaller Rolls-Royce turboshaft engine installed in the aircraft. Honeywell will supply the generator and Siemens the motor. This demonstrator will be an early step in Airbus's larger e-Thrust program. This will be a serial hybrid design where the gas turbine is used to generate electric power that then powers the six motors driving the propulsors. In the United States, NASA and its partners are teaming up to design a hybrid-electric aircraft that employs a parallel hybrid architecture where the underwing gas turbines provide both thrust as well as electrical power to drive an additional electric ducted fan in the rear. In both the proposed designs, the electric fans employ BLI concepts.

Materials and Manufacturing

Ceramics are a key to higher-temperature components in engines and consequently higher efficiencies. The new CFM LEAP engine uses silicon carbide-based ceramic matrix composites (CMC) in its hot section, which allows it to achieve higher temperatures and consequently higher cycle efficiencies. This allows the LEAP engine to rival the Geared Turbofan (GTF™) engine from P&W in fuel burn savings for the new Airbus A320neo family of aircraft [21].

FIGURE 22 Joby Aviation S2 AE Air Taxi.

FIGURE 23 Airbus A3 Vahana AE Air Taxi.

FIGURE 24 Hybrid-electric region aircraft concept.

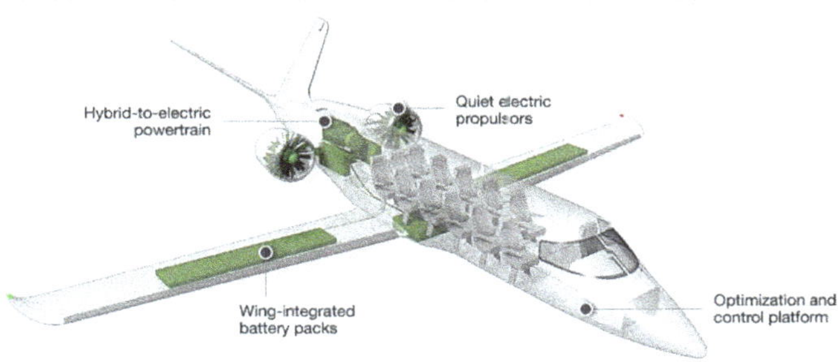

The trend to increase the percentage of high-temperature material in the engine core will continue to grow. Materials that can withstand higher temperatures need less cooling air, and this helps increase efficiency of the turbine.

Other composites are also being used in many areas of the engine to reduce weight. GE pioneered the use of composite fan blades on its GE90 engine. Composites have found their way into the nacelles, ducts, fan cases, fixed vanes, and other structures.

It is a similar story for the airframe, where weight is a crucial design parameter that is constantly under pressure to be reduced. Many GA aircraft manufacturers make most of their aircraft out of composites, but even larger commercial airframes are increasing the percentage of composites in their aircraft, even in load-bearing areas. Both Airbus and Boeing report that their latest aircraft are 50% or more composite by weight [22], and this fraction is only going to increase with future aircraft designs.

A major revolution in manufacturing is 3D printing, which will continue to grow to fill the needs of the aerospace industry. It is predicted that in the next 5 years or so, the global aerospace 3D printing market will grow on average by more than 55% each year [23]. Many noncritical airframe parts have been approved for use on aircraft, mainly inside the cabin. In 2017, Boeing and Norsk Titanium got Federal Aviation Administration (FAA) approval for a structural part for the B787 [24]. GE has been making thermocouple housings and fuel nozzles for engines using an additive process for a while now [25]. As OEMs gain more experience with these parts in the field, their development and acceptance will grow exponentially.

Smart Aircraft

While the aircraft as we know it will look and operate differently in the future, some of the most innovative improvements in aviation will not be immediately obvious to the eye. As with every other aspect of our lives, the aircraft is already getting more connected with the public and the private Internet. This trend is going to continue, and with it will come some revolutionary new things.

We are already seeing the more connected aircraft. In-flight connectivity has expanded in-flight entertainment (IFE) systems to provide real-time connectivity to the Internet and is being referred to as In-Flight Entertainment and Connectivity or Communication (IFEC). This is becoming standard on larger passenger aircraft, with satellite links allowing overwater broadband communications. Some airlines even offer in-flight mobile connectivity. While costs are still exorbitant, it is expected that this will come down as competition increases. Connection bandwidths are limited, and speeds are low, but this too is slated to increase as companies work to add communication infrastructure on the ground and in space [26]. What has not happened as readily is a broader growth in communications for technical data. The most advanced civil aircraft today, like the Boeing B787, generate about 100MB of data per hour. While popular magazine articles have quoted much higher numbers [27], this is not borne out by interviews with engineers at Boeing and Airbus. Nevertheless, with a more instrumented vehicle in the future, it is expected that this number will grow significantly, even if it does not reach the terabyte-per-flight level.

Today, there is no cost-effective way of transmitting, storing, and analyzing this data from each aircraft in a large airline fleet, especially during flight. With the advent of smarter asset health management technologies, this must change. Fast, reliable, and cost-effective data communications must become the norm. Satellite uplinks, with low-earth communication satellites, is an answer, but possibly a better answer would be to manage data communications in a more segregated and smarted manner. Critical messages that can affect the current flight should be provided a different communication channel, while noncritical data should be queued up for intermittent transmission over multiple channels. This could include satellites as well as ground-links, Wi-Fi at the gate, and cellular broadband communications. All these technologies are already available on modern aircraft. The challenges are more in the area of business models that will ensure that the data can be reliably stored and retrieved and analyzed continuously to provide maintenance actions to the operators. Data storage is another big challenge, and it is probably not hard to predict that cloud-based solutions will be very popular in the future. Data communication speeds will increase as well. Current Internet speeds are well under 1 Gb/s and much lower in much of the world outside the developed countries. The Global Lightpath e-Infrastructure (GLIF) has several 10 Gb/s networks in place in North America and Europe as part of a research network for academia and research organizations. Sections of this network are slated to run at 100 Gb/s speeds as well. This research network is slowly being made available to commercial users as well [28], so it is not inconceivable that such a network can be used in the future for transmitting full-flight engineering data from fleets of aircraft at a fraction of the time it takes today. This does not solve the issues related to slow wireless communication speeds, but because, in this case, we are talking about individual aircraft, the problem is not as acute.

Getting ready access to data is only the beginning of the story. The big breakthrough in the future will be in automated data analysis and decision-making. Artificial intelligence (AI) is gaining ground in other data-rich environments, and the aviation sector is bound to benefit from this growth. The vision for the smart aerospace supply chain has always been the seamless integration of all aspects of the maintenance, repair, and overhaul (MRO) supply chain. Data from each aircraft is analyzed against its own model (the so-called digital twin, which is not limited to the aerospace field alone [29]), and deviations or trends from the anticipated behavior are automatically scrutinized for signs of faults or impending failures. A positive diagnosis of a problem triggers a work order that is automatically scheduled to be performed at the most appropriate time at the correct location. For instance, the identification of a critical fault might entail an emergency repair being scheduled during the very next stop; with minimal disruption to revenue flights. A fault with a longer lead time might allow the plane to be routed to a more central overhaul facility, again with one of the optimization objectives being the minimization of disruptions to normal flight schedules. Pre-positioning of parts and optimal stocking of spares' inventories will be part of the optimal strategy.

Economic Models

Almost as important as the technological advances are the novel financial models that will be in place to operate the aviation marketplace. People are taking their cue from innovations in other sectors such as automobiles and proposing novel concepts. Take, for example, Uber-like air-taxi services. In fact, Uber itself has teamed up with aerospace companies to study the viability of such a service. It is in talks with Bell, Pipistrel, and even Embraer to develop an airplane that can be used for air-taxi service as early as 2020 in metropolises like Dallas-Fort Worth and Dubai [30]. As stated before, Airbus is working on its own air-taxi venture via a Silicon Valley subsidiary called A^3 (A-cubed). The autonomous vehicle they are building (Vahana) is designed to be summoned by an app on the smartphone to provide point- to- point service between designated locations. A number of other companies are building autonomous vehicles with similar air-taxi operations in mind. Volocopter in Germany and EHang in China are two examples, as is Kitty Hawk which recently revealed the Cora VTOL aircraft in New Zealand.

The automobile sector is going through its own rethinking of the existing model of ownership and usage. The current model is unsustainable in the long run, not just because of its reliance on fossil fuels but also because most light vehicles are simply parked and not doing anything for most of their lives. It has been estimated that personal cars in the United States are used only 5% of the time [31]. Hence, the same service of human transportation can be provided by a fraction of the number of vehicles if they are operating all the time. Of course, this reasoning is based on jumping to a drastic conclusion and does not take into consideration all the other uses an automobile is put to, but the fact remains that automobile usage today is very wasteful and some rationalization is sorely needed. Making any changes, of course, would require people giving up car ownership and just renting as needed, summoning it autonomously to the origin of the journey and dropping it off at the destination, whereby it returns to the closest depot autonomously. As more and more autonomous vehicles enter the roadways, it will soon become more dangerous to have (unpredictable) human operators intermixed with them; and a day may come when nonautonomous driving will become illegal in urban areas. This same model could well work for the aviation sector too. However, all this would require significant improvements to the air traffic control (ATC) system. Even today, as more drones are entering the airspace, it is far from certain that the current ATC is up to the task of handling all this traffic safely. When autonomous human-carrying aircraft start flying around, something far more advanced than even a next-generation ATC will be needed.

Aircraft ownership and operation is another aspect of the aviation business that might change drastically in the future. Even today in civil aviation, the owner of the aircraft is often a leasing company and not the entity operating the asset. In 2017, the two biggest owners of aircraft in the world were AerCap (1,153 aircraft) and GE Commercial Aircraft Services (931) [32]. By contrast, the world's largest airline fleet—American Airlines—owned 950 aircraft. Many airlines prefer to lease aircraft rather than own them and to spend their money on flight operations instead. In fact, many airlines don't even maintain their own aircraft, trusting this job to third-party maintenance companies or the maintenance operations of major airlines. It is possible that this trend toward specialization will continue in the future, with more and more airlines concentrating their resources on what they know best, that is, transporting passengers and cargo, and sourcing out the rest of the functions to companies specializing in other aspects of the aviation business.

As we look to the future, we are going to see some drastic changes to the way the aviation business is conducted around the world. Some of these will be brought about by technological advances and some by changes in business models. While many changes are predictable, as we have outlined above, it is also safe to say that there inevitably will be new concepts and designs that will appear from seemingly nowhere. Never forget that we humans have the bottomless capacity to surprise ourselves with our ingenuity and inventiveness.

References

1. Federal Aviation Administration, "FAA Aerospace Forecast: Fiscal Years 2017-2037," Retrieved July 2017, https://www.faa.gov/data_research/aviation/aerospace_forecasts/media/FY2017-37_FAA_Aerospace_Forecast.pdf.
2. Air Transport Action Group, "Facts & Figures," accessed February 2018, https://www.atag.org/facts-figures.html.
3. Yutko, B. and Hansman, J., "Approaches to Representing Aircraft Fuel Efficiency Performance for the Purpose of a Commercial Aircraft Certification Standard," MIT International Center for Air Transportation, Report ICAT-2011-05, May 2011.
4. Aviation Safety Network Database, "Flight Safety Foundation, Accident List: de Havilland DH-106 Comet," accessed February 2018, http://aviation-safety.net/database/types/de-Havilland-DH-106-Comet/losses.
5. Boeing 737 MAX 7 Technical Specs, "Boeing Commercial Aircraft," accessed January 2018, http://www.boeing.com/commercial/737max/.
6. Airbus A319neo, "Airbus," accessed January 2018, http://www.airbus.com/aircraft/passenger-aircraft/a320-family/a319neo.html.
7. Kharina, A. and Rutherford, D. "Fuel Efficiency Trends for New Commercial Jet Aircraft: 1960 to 2014," White paper, The International Council on Clean Transportation, August 2015.
8. General Aviation Manufacturers Association, "2016 General Aviation Statistical Databook & 2017 Industry Outlook," Retrieved June 2017, https://gama.aero/wp-content/uploads/2016-GAMA-Databook_forWeb.pdf.
9. Federal Aviation Administration, "Fact Sheet – General Aviation Safety," October 24, 2017, Retrieved 5 February 2018, https://www.faa.gov/news/fact_sheets/news_story.cfm?newsId=21274.
10. Boeing, "Current Market Outlook 2017-2036," Retrieved January 2018, http://www.boeing.com/commercial/market/current-market-outlook-2017/.
11. IATA, "Fact Sheet, Industry Statistics," December 2017, Retrieved May 2018, http://www.iata.org/pressroom/facts_figures/fact_sheets/Documents/fact-sheet-industry-facts.pdf.

12. IATA, "Fact Sheet, Alternative Fuels," December 2017, Retrieved May 2018, https://www.iata.org/pressroom/facts_figures/fact_sheets/Documents/fact-sheet-alternative-fuels.pdf.
13. Federal Aviation Administration, "Fact Sheet – Leaded Aviation Fuel and the Environment," June 2013, accessed January 2018, https://www.faa.gov/news/fact_sheets/news_story.cfm?newsId=14754&omniRss=fact_sheetsAoc&cid=103_F_S.
14. Trimble, S., *Analysis: Aerion Shows a Configuration of Supersonic Jet Engine*, Flight Global, Washington, February 22, 2018, https://www.flightglobal.com/news/articles/analysis-aerion-releases-configuration-of-supersoni-446122/.
15.. Ward, T.A., "*Aerospace Propulsion Systems*," (US: John Wiley & Sons, 2010.
16. Pearce, W., *Allison V-3420 24-Cylinder Aircraft Engine*," Old Machine Press, https://oldmachinepress.com/2017/04/20/allison-v-3420-24-cylinder-aircraft-engine/, image downloaded April 2016.
17. UPS Battery Center Ltd., "Robert Anderson (19th century Scottish Inventor)," May 2014, accessed January 2018, http://www.upsbatterycenter.com/blog/robert-anderson-19th-century-scottish-inventor/.
18. Rajamani, R., "*Electric Flight Technology: Unfolding of a New Future*," (Warrendale: SAE International Press, April 2018.
19. Gallagher, K.G., Goebel, S., Greszler, T., Mathias, M. et al., "Quantifying the Promise of Lithium-Air Batteries for Electric Vehicles," *Energy & Environmental Science*, 5, 2014, 1555-1563.
20. Bellamy, W., "*Airbus E-Fan X First Flight to Occur in 2020*," (USA: Avionics, November 2017, Retrieved April 2018, http://www.aviationtoday.com/2017/11/28/airbus-e-fan-x-first-flight-occur-2020/.
21. Levy, D. "Ceramic Matrix Composites Take Flight in LEAP Jet Engine," January 2017, Retrieved July 2017, https://phys.org/news/2017-01-ceramic-matrix-composites-flight-jet.html.
22. Chady, T., "Airbus Versus Boeing – Composite Materials: The Sky's the Limit," LeMauricien.com, September 2013, Retrieved July 2017, http://www.lemauricien.com/article/airbus-versus-boeing-composite-materials-sky-s-limit.
23. Kerns, J., "*Commercial Supersonic Flights No Longer Pie in the Sky*," (Washington: Machine Design, July 2017, Retrieved July 2017, http://www.machinedesign.com/3d-printing/commercial-supersonic-flights-no-longer-pie-sky.
24. Scott, A., "*Printed Titanium Parts Expected to Save Millions in Boeing Dreamliner Costs*," (London: Reuters, April 2017.
25. Kellner, T. "The FAA Cleared the First 3D Printed Part to Fly in a Commercial Jet Engine from GE," GE Report, April 2015, Retrieved July 2017, http://www.ge.com/reports/post/116402870270/the-faa-cleared-the-first-3d-printed-part-to-fly/.
26. Moores, V., "*Inmarsat: Wi-Fi is a Necessity, Not a Luxury*," (USA: Air Transport World, July 2017, Retrieved August 2017, http://atwonline.com/ife/inmarsat-wi-fi-necessity-not-luxury.
27. Finnegan, M., "*Boeing 787s to Create Half a Terabyte of Data per Flight, Says Virgin Atlantic*," Computer World, UK, March 2013.
28. Gommans, L., "Sharing Big Data Asset via a Secure Digital Market Place," *SAE HM-1 Meeting*, Fort Worth, TX, March 2017.
29. Marr, B., "*What Is Digital Twin Technology – And Why Is It So Important?*" Forbes Tech, March 2017, Retrieved July 2017, https://www.forbes.com/sites/bernardmarr/2017/03/06/what-is-digital-twin-technology-and-why-is-it-so-important/.

30. Hawkins, A.J., "Uber is Getting Serious about Building Real, Honest-to-God Flying Taxis," *The Verge*, April 2017, Retrieved May 2017, https://www.theverge.com/2017/4/25/15417506/uber-flying-taxi-elevate-dallas-dubai-bell-helicopter.

31. Neil, D., "Could Self-Driving Cars Spell the End of Ownership?" *Wall Street Journal*, December 2015, Retrieved July 2017, https://www.wsj.com/articles/could-self-driving-cars-spell-the-end-of-ownership-1448986572.

32. Morris, R., "Commercial Operating Leasing Market Dynamics," Flight Ascend Consultancy Viewpoint, 55, 2nd quarter 2017, downloaded February 2018, https://www.flightglobal.com/asset/16524.

appendix

Abbreviations

AC - Advisory Circular
AD - Airworthiness Directive
AHM - aircraft health management, aircraft health monitoring
AM - additive manufacturing
ARINC - Aeronautical Information Incorporated
ARP - aerospace recommended practice
ASM - available seat mile
ATA - Air Transport Association
ATC - air traffic control
ATM - air traffic management
CAA - Civil Aviation Authority
CAD/CAM - computer-aided design and computer-aided manufacturing
CAR - civil air regulation
CASM - cost per available seat mile
CDC - Centers for Disease Control
CEO - conventional engine option
CFD - computational fluid dynamics
D&C - delayed or canceled
DOC - direct operating costs
EASA - European Aviation Safety Agency
E-bay - electronics bay
EHM - engine health management
FAA - Federal Aviation Administration
FADEC - full authority digital engine controller
FAR - Federal Aviation Regulations
FBW - fly-by-wire
FDA - Food and Drug Administration
FDALs - functional development assurance levels
FHA - functional hazard assessment
FMEA - failure modes effects analysis
FMECA - failure modes, effects, and criticality analysis
FTA - fault tree analysis
GAMA - General Aviation Manufacturers Association
GDP - gross domestic product
GHG - greenhouse gas
HP - high-pressure
IATA - International Air Transport Association

ICAO - International Civil Aviation Organization
IDALs - item development assurance levels
IT - information technology
IVHM - integrated vehicle health management
JAA - Joint Aviation Authorities
LCC - low-cost carrier
LP - low pressure
MMI - man-machine interface
MPIG - Maintenance Program Industry Group
MRB - Maintenance Review Board
MRO - maintenance, repair, and overhaul
MSG - Maintenance Steering Group
NASA - National Aeronautics and Space Agency
NATO - North Atlantic Treaty Organization
NEO - new engine option
OEM - original equipment manufacturer
OSHA - Occupational Safety and Health Administration
PC - production certificate
PMA - Parts Manufacture Authority
P&W - Pratt and Whitney
RASM - revenue per available seat mile
RCM - reliability-centered maintenance
R&D - research and development
RTCA - Radio Technical Commission for Aeronautics
SST - supersonic transport
STC - supplemental type certificate
TC - type certificate
TLRs - top-level requirements
TSA - Transportation Security Administration
TSO - Technical Standard Order
UA&TC - United Aircraft and Transportation Company
UAS - unmanned aircraft systems

about the authors

David Alexander has been with SAE International and its affiliates since 2004 and is based in London, UK. In May 2016, he assumed the role of Director, Aerospace Standards. David is responsible globally for the operation of the SAE Aerospace Standards program comprising of over 7,500 industry standards maintained by more than 180 technical committees. This includes the management support for the SAE Aerospace Council and SAE Executive Standards Committee, and involves leveraging standards to work across the SAE International portfolio.

He is supported by SAE staff in London, Paris, Brussels, Shanghai, and across the United States, including relationships with industry, associations, aviation authorities, and government stakeholders around the world.

David Alexander maintains the SAE focal representation with EASA, and coordinates the aerospace and new standards development activities within the SAE Aerospace Standards team, having previously established the strategic IVHM and Electric Aircraft Steering Groups with industry leaders. He also facilitates the General Assembly of the International Aerospace Quality Group (IAQG).

Prior to joining the SAE Aerospace Standards operation, David worked on the Nadcap accreditation program through SAE's affiliate organization PRI, and he holds a BA (Hons) degree from the University of Manchester.

Professor John Fielding started his career as an apprentice and design engineer in Manchester, working on the Nimrod, BAE 748, and BAE 146 aircraft. After twelve years, he joined Cranfield University as the first Air Registration Board Research Fellow, with subsequent promotions leading to becoming Professor of Aircraft Design and Head of the Department of Aerospace Engineering. Concurrently to his academic endeavors, during the period 2005-2010, John was also Chief Engineer of the BAE SYSTEMS/EPSRC-sponsored DEMON technology demonstrator UAV.

Although partially retired, John continues as a Professor Emeritus at Cranfield University, where he teaches and carries out various research projects. He is the founder and director of Fielding Aerospace Consultants Ltd. and specializes in research and education in aircraft design, reliability, maintainability, and operational effectiveness.

He acts as a consultant and expert witness and has published more than 100 technical papers at conferences and in journals. His popular textbook *Introduction to Aircraft Design*, published by Cambridge University Press, is now in its second edition.

John Fielding is a member of SAMAD Aerospace's advisory board, and chairman of the Cranfield College of Aeronautics Alumni Association.

Richard Gardner, MCIPR, MRAES started his career with British Airports Authority in the planning department, then became public relations executive, organizing media relations and corporate exhibitions and events. He joined the UK Ministry of Defence in 1973 and spent 11 years in Royal Air Force Public Relations in the London MOD Headquarters Press Office and later writing and editing official publications. He served for three years in RAF Germany and later became Joint Services Public Relations Officer South Atlantic, based in the Falklands.

Richard left MOD in 1984 and became a PR Director in a City consultancy in London, managing media relations teams for UK aerospace and defense companies and ran Airbus corporate UK PR campaign for 6 years.

He became an independent consultant and writer in 1992 promoting new developments, products, and services for clients in the aerospace and defense hi-tech domain. He edited *Aerospace* magazine for the Royal Aeronautical Society for 10 years, 2001-2010, and was invited to write the society's 150th Anniversary commemorative book in 2016, and currently edits the Air League Newsletter. Now retired from public relations, he continues to write on a freelance basis as an author on aviation and defense matters and as a TV commentator and consultant on aviation history. He regularly contributes to SAE Aerospace reports.

A trustee of Farnborough Air Sciences Trust since 1994, Richard campaigned successfully to save the historic wind tunnels and aviation buildings at Farnborough. He is Chairman of the Trust, a not-for-profit charity, and is responsible for the FAST Museum dedicated in safeguarding the heritage of the nation's oldest aviation research and development center, with an archive including hundreds of thousands of technical reports, images, films, artifacts, prototype equipment, missiles, satellites, helicopters, UAVs, and fast jet aircraft. His museum partners include the (UK) National Archives, Hampshire County Council, and the Science Museum.

Richard Greaves graduated from University College London in 1968 with a degree in Physics. He went to work in the United Kingdom Atomic Energy Authority in their Reactor Control and Instrumentation division at Winfrith Heath. His scholarship with the UKAEA enabled him to study part time at the University of Southampton for his Doctorate, awarded in 1973.

At the end of 1972, he started working for Vibro-Meter S.A. in Fribourg, Switzerland, heading their piezo-electric transducer engineering department. This serendipitous move took him out of the nuclear industry and into aerospace. Vibro-Meter became a major world player in aircraft gas turbine vibration monitoring. In 1981, Richard moved to Los Angeles to lead the company's operations in the US, and during that time, became a member of SAE International joining their E-32 Engine Health Monitoring Technical Committee.

In 1990, after leaving Vibro-Meter, Richard set up his own company, Greaves & Associates, which acted as sales representatives in Eastern Europe in the field of Engine Health Monitoring. From 1994, Richard lived in Moscow before returning to Switzerland in mid-1996 as President of Vibro-Meter S.A., which was acquired in 1998 by Meggitt Plc. Richard then was made the divisional head of Meggitt Sensing Systems responsible for nine companies in six countries with about $500M in sales. In 2009, he was promoted to the position of Senior V.P. of Engineering and Technology and, in 2012, Chief Technology Officer Emeritus.

In 2009, Dr. Greaves was recognized as a Fellow of SAE International, and in 2013 he was awarded Fellowship at the Royal Academy of Engineering in the United Kingdom.

He served on SAE's Board of Directors from 2012 to 2016 and was its President and Chairman in 2015.

Christer Hellstrand is an engineer and statistician, and has held multiple senior positions in the aerospace industry, such as Director of Quality Management, Operations, and Supply Chain Management at the Boeing Company.

Prior to this position, he also worked for SKF, a global bearing manufacturer. In 2011, Christer spearheaded the launch of the International Environmental Group (IAEG) as the director of EHS for the Boeing Company. He led the IAEG Board as its chairman until his retirement from Boeing in 2015. Since then he has continued to support IAEG as its Program Director. In addition to the United States, Christer has lived and worked in Sweden, the Netherlands, UK, Malaysia, and Japan. He is currently the principal of Christer Hellstrand Consulting, based out of Seattle, WA.

Ian K. Jennions is a professor and director of the IVHM Centre, Cranfield University, UK. He joined the Centre, which is funded by a number of industrial partners, when it was founded in 2008 and has led its development and growth in research and education over the last 10 years. Previously, Ian had worked for a number of companies in the gas turbine industry over a 40-year career. He worked for Rolls-Royce, General Electric, and Alstom in a number of technical roles, gaining experience in aerodynamics, heat transfer, fluid systems, mechanical design, combustion, and, more recently, IVHM. He has a mechanical engineering degree and a PhD in CFD, both from Imperial College, London. He is a Director of the PHM Society, Chair of the SAE IVHM Steering Group and contributing member of the HM-1 IVHM committee, and a Fellow of IMechE, RAeS, and ASME. He is also the editor of five SAE books on IVHM and a coauthor of an SAE book on No Fault Found.

Hals Larsen has over 40 years of experience in the aerospace industry, first at the Boeing Company and later with the FAA. In his long career, Hals has worked on many aspects of aircraft propulsion and propulsion controls, and their integration with the airframe. Before retiring as the Director of Propulsion Electronics, Hals worked with every engine manufacturer on all large commercial jet programs including the B747, the B767 and the B787. During this period his focus was on engine controls and their interfaces with the airplane and with the cockpit displays. With the advent of electronic engine controls (EEC), Hals was critically involved with their certification and integration. Earlier in his career he was a key part of the auto-throttle work that was being introduced on the B747.

After retiring from Boeing, Hals took a job with the FAA as their Director of Propulsion Electronics. There he became familiar with the smaller aircraft and gave many FAA talks and attended many discussions of electronic control system integration with aircraft systems. He wrote documents on EEC systems, their integration, and their reliability. Over the years at the FAA, he developed relationships with his European counterparts and developed ways of synchronizing the two points of view. As such, the FAA job gave him a good opportunity to listen to and understand the outlooks of other companies and countries on propulsion electronic system "design and certification." Hals has a BS and an MS in Mechanical Engineering from Purdue University, majoring in control systems and thermodynamics. He has been involved with SAE International and their E-32 Aerospace Propulsion Systems Health Management technical committee for a number of years.

Steve O'Flarity is an engineer who has worked on the development and validation of military and civil gas-turbine aero engines throughout his career, spanning 38 years at Pratt & Whitney (P & W). He led developmental gas turbine and aircraft test activities at ground, simulated altitude, and flight test facilities for half that period. He was program manager for the development of P&W's Advanced Diagnostics and Engine Management system, which utilizes flight data to perform prognostic and diagnostic analyses of in-service civil and military engines. Steve defined and led the development of P&W's Flight Data Analysis, Storage, and Transmission system to seamlessly collect transient flight data from engine start to landing, and deliver it to users. He served as deputy program manager for the development and certification of the revolutionary Geared Turbofan™ engine's initial application. He was the executive technical assistant to one of the world's most awarded aerospace engineers, Dr. Alan Epstein, while also leading teams conducting advanced design and analysis of distributed propulsion systems, extremely small high-performance engines, and hybrid/electric aero propulsion, with patents granted and pending. Steve has been a long-time active member of SAE's E-32 Aerospace Propulsion Systems Health Management committee and the Integrated Vehicle Health Management Steering Group, as well as the AIAA and ASTM technical committees.

He holds a BS degree in mechanical engineering, an ME in industrial and systems engineering, and an MBA from the University of Florida, as well as an MA in sustainability and environmental management from Harvard University.

David Piotrowski currently works as a Principal Engineer in Delta Air Lines' Enabling Technologies group. His responsibilities include innovation of inspection technologies, oversight of Delta's internal NDT materials and processes, developing and implementing novel inspections to solve in-service problems, and assisting in Damage Tolerance Analyses and Failure Investigations.

Prior to his 18 years at Delta, David worked at Pratt & Whitney for 4½ years in a similar capacity. Growing up in an airline family (father was air line mechanic, mother a flight attendant) provided David with unique perspectives of the industry, spanning decades.

He was part of teams which were named the 2006, 2008, 2009, 2010, 2012, and 2014 FAA-A4A "Better Way Award" winners for accomplishments in NDT. In 2010, he received Delta's highest honor for dedication and determination, being inducted into the "Chairman's Club."

He has Bachelor of Materials Engineering and Master of Metallurgy degrees from Georgia Tech, as well as a Master of Aeronautical Science (specializing in Technical Management) from Embry-Riddle University. He is an American Society for Nondestructive Testing (ASNT) Level III certified in penetrant, ultrasonics, and eddy current.

He resides in Atlanta, GA (USA), and is married to his wife Stephanie. They have three wonderful children who provide plenty of excitement and even more interesting times.

Ravi Rajamani is an independent consultant who has been working for many years in the area of aerospace propulsion and energy, specifically in data analytics and model-based methods for controls and diagnostics. He has many publications to his name including book chapters, journal and conference papers, and patents. Prior to his current job, Ravi worked at Meggitt, United Technologies Corporation, and General Electric Company. Dr. Rajamani has a BTech (ME) from IIT Delhi, an MS (Automation) from IISc, Bangalore, an MBA from the University of Connecticut, and a PhD (EE) from the University of Minnesota. He is active within SAE's Propulsion Health Management (E-32) and Integrated Vehicle Health Management (HM-1) committees; and its Electric Aircraft, Digital & Data, and IVHM Steering Groups. He is also active in the PHM Society, serving on its board of directors, and he was the general chair of the 2014 European PHM conference in Nantes, France. Ravi is a Visiting Professor of Aerospace, Transport, and Manufacturing at Cranfield University in Cranfield, UK. He has been elected a fellow of SAE and IMechE and is a recipient of the SAE Forest R. McFarland Award. He also serves as a board member of the Edison Tech Center in Schenectady, NY.

Robert Voros is the Regulatory Compliance Leader—Development Assurance & System Safety for Textron Aviation Inc. Textron Aviation Inc. is home to the Beechcraft, Cessna, and Hawker brands. Its products range from small airplanes, Part 23, to transport category airplanes, Part 25, including Citation business jets, King Air and Caravan turboprops, and T-6 military trainer aircraft. As a regulatory compliance leader, Robert is the primary technical adviser to the Textron Aviation Organization Designation Authorization, regarding development assurance and system safety. He is also a key interface on these topics to both industry organizations and certification authorities at the local and policy levels.

With more than 20 years of experience, Robert has held a variety of positions of increasing responsibilities, starting at Cessna Aircraft Company as a flight control system design engineer. He has supported the type certifications and system safety assessments of multiple aircraft for new and amended type certificates. In this time, he has also achieved a Design for Six Sigma Black Belt, 2 patents, and 5 published papers. Robert earned a Bachelor of Science degree in Mechanical Engineering from Rose-Hulman Institute of Technology.

Robert is currently serving as the Chairperson for SAE International's S-18 Aircraft and System Development and Safety Assessment Committee. The S-18 committee brings together qualified specialists for the advancement of aerospace safety and to support effective safety management through the development of technical reports that are key to the civil certification of aircraft.

index

Acceptable means of compliance (AMC), 155, 156
Acclimatization, 91
Acquisition cost, 34, 35, 86, 163, 174
Additive manufacturing (AM) techniques, 51
Advancing Blade Concept (ABC), 174
Advisory circulars (AC), 71
Advisory Council for Aeronautical Research in Europe (ACARE), 38
Advisory Group for Aerospace Research and Development (AGARD), 64
Aerion AS2 aircraft, 170
Aerodynamics, 63
  shapes, 34
  testing, 58
Aeronautical Information Incorporated (ARINC), 149
Aerospace Industry Steering Committee on Structural Health Monitoring, 153
Aerospace Recommended Practices (ARP)
  ARP4761, 71
  ARP4754A, 71
Aerospace value chain, 5
Aft fuselage-mounted engines, 30, 32
Aft-swept wings, 30
Agência Nacional de Aviação Civil (ANAC), 103
Airbus A350, 51, 62
Airbus A350 composite wing structural test, 60
Airbus A380, 19, 27, 32, 37, 51
Airbus A319neo, 163
Air Commerce Act, 44
Aircraft design
  airliner categories
    medium/large wide-body airliners, 27
    regional airliners, 26
    single-aisle airliners, 26
    small wide-body airliners, 26
  and development process
    aircraft requirements capture, 27–30
    airlines, 32
    CAD/CAM systems, 32
    classic life cycle, 27
    collaboration and integration, 32
    conceptual design, initial layout, 30–32
    design process, 27, 28
    funding issues, 32
    global participation, 32
    high-subsonic airliners, 32
    MSG-3 process, 32
    OEMs, 32
  meet/exceed passenger expectations
    comfort and communications, 37–38
    DOC elements, 33–37
    environmental responsibility, 38
    safety, 33
    speed and flight frequencies, 38
  passenger expectations, 26
  technologies and configurations, 38–39
Aircraft development, 3
Aircraft functional hazard assessment, 121–123
Aircraft health management, 21
Aircraft health monitoring, 107
Aircraft load agents, 90
Aircraft noise
  day-to-day operational procedures, 144
  ICAO aircraft noise strategy, 143
  ICAO noise standards, progression of, 143
  land use, planning and management, 144
  nacelle design, 144
  operational restrictions, 144
Aircraft operation and maintenance in service, 3
Aircraft ownership and operation, 182
Aircraft requirements capture
  aircraft top level requirements, 30
  customer needs, air transport market research, 28–29
  safety and environmental requirements, 29–30
Aircraft safety assessment, 127
Airframe
  aft BLI concept, 175, 176
  Airbus A3 Vahana AE Air Taxi, 179
  air traffic management, 178
  Aurora Double-Bubble/BLI concept, 175
  B-2 flying wing, 175, 176
  boom supersonic jet, 174
  EHAs, 176
  EMAs, 177
  Eurocopter X3, 175
  hybrid-electric region aircraft, 178, 179
  Joby Aviation S2 AE Air Taxi, 179
  Kitty Hawk Cora AE Air Taxi, 178
  light aircraft space, 177
  NASA HWB concept, 176
  NASA X-48 BWB, 176
  RACER, 175
  Sikorsky X2, 174, 175
  turboprop-powered aircraft, 177
  urban mobility, 177
Airline accidents, 2
Airliner cruise speeds, 38
Airliner list prices, 34, 35
Air Mail Act, 44
Airmail aircraft, 13
Airplane Health Management (AHM), 110
Airport customer service, 89–90
Air traffic control (ATC), 138, 155, 182
Air traffic management (ATM), 2, 150, 178
Air Transport Association (ATA), 18
Air transport market research, 28–29
Airworthiness directives (ADs), 73, 106, 155
Airworthiness limitation items (ALIs), 106
Allison V-3420-B-5 engine, 171
Altimeter, 10
America Makes and ANSI Additive Manufacturing Standardization Collaborative, 154
American Airlines, 83, 87
American Society for Testing and Materials (ASTM), 140, 150, 152, 155, 157
Anglo-French Concorde supersonic airliner, 47
APB Winglets Co., 109
Artificial intelligence (AI), 181
ASD-CERT program, 152

Association for Safe International Road Travel, 2
ASTM International, *see* American Society for Testing and Materials (ASTM)
Atlas, 85
Autoclaves, 52
Automated bag check-in, 3
Automated composite wing spar jigs, 50
Auxiliary power unit (APU), 177
Available seat miles (ASMs), 28, 109
Aviation Rulemaking Committee (ARC), 157
Aviation safety, 26, 29–30
Aviation Week, 140

"Baggage handlers," 90
"Basic Economy" fares, 88
"Belly capacity," 85
Big Data, 110
Biofuels
  life cycle, 140, 141
  second-generation feedstock for
    algae, 142
    camelina, 142
    halophytes, 142
    jatropha, 142
Biometrics, 90
Biplane glider, 10
Blade-off rig tests, 61
Blended wing body (BWB), 176
Boeing 707, 46
Boeing 737-700, 163
Boeing 737-800, 140
Boeing 747, 27, 140
Boeing 787, 32, 51, 65
Boeing Airplane Company, 13
Boeing Air Transport Company, 44
Boeing B787, 177
Boeing factory, 44
Borschberg, André, 140
Boundary layer ingestion (BLI), 171, 175, 176
British Airways, 83, 84, 86, 87
British Standards Whitworth, 148
Business strategy, operators
  carriers, 83
  fleet strategy and products, 85–87
  network planning, 83–85
  perished inventory, 83
  public service, 83
  route networks, 83
Bypass ratio (BPR), 166, 168–170
Carbon Offset and Reduction Scheme for International Aviation (CORSIA), 139
CareServices, 108
Cargo, 47, 49, 51, 82, 83, 85, 92, 102, 110, 140, 144, 162, 163, 173, 176, 183
Carriers, 83
CATIA digital tools, 49
Cayley, George, 10
Ceramic matrix composites (CMC), 179
Certification process
  airworthiness certificate, 73
  ARP4761, 71
  ARP4754A, 71
  engine, 75–78
  failure modes, 72
  goal of, 72
  Maintenance Review Board process flowchart, 73, 74
  production certificate, 70, 71
  RTCA, 71
  tests, safety
    analytical techniques, 74
    engine-out takeoff tests, 75
    flight test, 74
    functional validation, 74
    system verification, 74
  type certificate, 70, 71
  typical product development cycle, 72, 73
Certification Specifications (CS), 155
CFM LEAP engine, 179
Changi Airport, 96
Civil aerospace
  aerial war, 13
  aerodynamic control, 11
  airframe, 173–179
  air rearwards, 17
  A320neo, 165
  commercial air transport management, 21
  C Series, 20
  economic models, 182–183
  E Series, 20
  fixed-wing aircraft development, 11
  floating lanterns, 9
  flying machines, 10
  fuel-efficient engines, 21
  integrated support entity, 22
  internal combustion engine, 14
  jet engine, 15
  lift and gravity forces, 11
  materials and manufacturing, 179–180
  MSG, 18
  piston engine, 15
  piston engine DC aircraft, 17
  production license, 16
  propulsion, 166–173
  radial engines, 15
  reconnaissance and bombing, 12
  rigid airships, 11
  smart aircraft, 180–181
  standards (*see* Standards)
  stationary crankshaft, 14
  synchronization gear, 12
  3D printing, 180
  turbojet engine, 17
  two-spool jet engine, 17
  wide-body civil transport aircraft, 18
Civil aerospace sector, 4, 134, 158
Civil Air Regulation (CAR), 117
Civil aviation, 6
  acquisition costs, 163
  emissions costs, 163
  factors, 162
  general aviation, 165
  MIL-P-1629, 117
  noise costs, 163
  operating costs, 163
  operating revenue, 163
Civil Aviation Administration of China (CAAC), 103
Civil Aviation Authority (CAA), 103
Civil Aviation Authority of Singapore (CAAS), 155
Civil Aviation Safety Authority (CASA), 103
Civil Standards Campaign, 155
Climate change
  Action on Climate Change, 134, 135
  Paris Climate Agreement, 134
  3rd Aviation and Environment Summit, 134
  UN Climate Conference in Copenhagen, 134
  UN's Intergovernmental Panel on Climate Change report, 134
Code of Federal Regulations (CFRs), 70, 74, 103, 107, 109
"Code-shares," 86
Comet aircraft fuselage, 59
Comfort, 26
Commercial aircraft market predictions, 28, 29

Commercial airline flights, 2
Common cause analyses, 121
  common mode analysis, 127
  particular risk analysis, 127
  zonal safety analysis, 126
Common mode analysis, 127
Communications, 26
Compensation, 3
Complex machinery, 60–61
Composite fuselage, 60
Computational fluid dynamics (CFD), 27, 33, 64
Computer-aided design (CAD), 27, 32, 35, 49, 52
Computer-aided manufacturing (CAM), 32, 35, 49, 52
Concorde, 18, 19, 169, 170
Concorde supersonic airliner, 38
Condition-based maintenance (CBM), 106
Continued operational safety (COS), 128, 129
Convair 880, 17
Convention on International Civil Aviation, 116
Corporate average fuel economy (CAFE), 171
Corporate safety and security, 93
Corrosion prevention and control programs (CPCP), 106
Cost per available seat mile (CASM), 87
Cranfield 152-seat airliner, conceptual design, 30
Crew costs, 36–37
Crew scheduling, 88, 90, 91, 93, 95
Curtiss Aeroplane, 12
Customer service agents, 90

De Havilland Comet, 16, 17, 59
De Havilland Comet 1, 163, 164
Delayed/canceled (D&C) flights, 3
Delta Air Lines, 83
Delta TechOps, 109, 110
Depreciation costs, 34–35
Design Assurance Guidance for Airborne Electronic Hardware, 125
Destructive testing, 59, 61
Development costs, 34–35
DH Comet jetliner, 46, 59
DHL, 85
"Digital Avionics Software," 118
Digital cockpit, 54
Digitally controlled drilling, 50
Digital twin, 181
Direct operating cost (DOC) reductions, 30
  crew costs, 36–37
  depreciation costs, 34–35
  fuel and oil costs, CO_2 reductions, 34
  maintenance costs, 35–36
Directorate General of Civil Aviation (DGCA), 103
Dispatchers, 92
Distributed propulsion, 171
Diverted flight, 94
Doors, 31
Douglas DC-1, 45
Douglas DC-2, 45
Douglas DC-7, 46
Douglas DC-8, 17, 46

EasyJet, 83, 87
E-Fan X, 179
Electrohydrostatic actuators (EHAs), 176
Electromagnetic pulse tests, 61
Electromechanical actuators (EMAs), 177
Electronics bay (e-bay), 75
Embraer E series, 20
Emergency exits, 31
Emirates Airlines, 87
Emission reduction
  biofuels
    life cycle, 140, 141
    second-generation feedstock for, 142
  economic measures, 139
  fossil fuel, 140, 141
  fuel efficiency and aircraft design
    aerodynamic improvements, 136
    aircraft energy intensity, 137
    designing efficient airplanes, 136
    new commercial jet aircraft, fuel efficiency trends for, 136, 137
  IATA road map, 136
  operations and infrastructure, 138
Employees, 94–95
1-40 (J33) engine, 16
Engineering process, 3
Engineering Standards Committee, 148
Environment
  aircraft noise, 142–144
  climate change, 134–136
  emission reduction, 136–142
Environmental responsibilities, 26
Environmental tests, 61
EUROCAE, 155
Eurocopter X3, 175
European Airbus consortium, 48
European ATM Standards Coordination Group, 154
European Aviation Safety Agency (EASA), 58, 70, 103, 120, 148, 155, 157
European Organisation for the Safety of Air Navigation (EUROCONTROL), 155
European Parliament and Council, 102
European regulation, 3
European UAS Standards Coordination Group, 154
European Union (EU), 3
Extended twin-engine operations (ETOPS), 106

Facial recognition, 90
Fail-safe design, 116
Failure modes and effects analysis (FMEA), 71, 119, 126
Failure modes, effects, and criticality analysis (FMECA), 117, 118, 126
Fan pressure ratio (FPR), 170
Fatigue testing, 59
Fault tree analysis (FTA), 71, 119, 126
Federal Aviation Administration (FAA), 70, 71, 73, 74, 77, 103, 106, 107, 117, 120, 138, 148, 155, 156, 180
Federal Aviation Regulations (FARs), 70, 92, 103, 117
FedEx, 85
Finances, operators
  CASM, 87
  "cost of capital," 87
  costs, 88
  fares and pricing, 88
  long-term stability, 87
  RASM, 87
  U.S. carriers, 87
  yield, 87
Finite element analyses (FEA), 33
Finite Element Structural Analysis, 27
Fleet strategy and products, 85–87
Flight
  frequencies, 26
  operations, 93
  tests, 64–66
Flightpath 2050, 2
Fly-by-wire (FBW) systems, 62, 63
Fokker Eindecker monoplane, 12
Forecasts, 2, 3
Four-engine aircraft, 14

Fuel and oil costs, CO_2 reductions, 34
Full authority digital engine controllers (FADEC), 76
Functional Development Assurance Levels (FDALs), 125
Functional Hazard Assessment (FHA), 71
Fuselage, 31, 37

Geared TurboFan (GTF), 20, 165–167, 179
General aviation (GA), 165, 168
General Aviation Manufacturers Association (GAMA), 157, 165
General Electric (GE) Aviation CFM56 engine core, 170
General Electric CF6, 140
Global Lightpath e-Infrastructure (GLIF), 181
Global Positioning System, 138
Global warming, 30
Greenhouse gas (GHG) emissions, 134
  emission reduction (*see* Emission reduction)
  high-level goals, 134
Gross domestic product (GDP), 1
Ground-based test rigs, 61
Ground support equipment (GSE), 36
Guidance Material (GM), 155
Guidelines and Methods for Conducting the Safety Assessment Process on Civil Airborne Systems and Equipment, 120

Haneda Airport, 86
Hawker Aircraft, 13
Hawker Hurricane fighter, 13
Hawker Siddeley Trident, 66
Heathrow Airport, 84
Heinkel H178 aircraft, 16
Heldt, Peter, 149
Helicopters, 14
Honeywell turbofan engines, 179
The Horseless Age, 149
Hot-air balloon flight, 10
"Hub-and-spoke" system, 83
Hybrid wing body (HWB), 176

Ilya Muromets type V, 14
Industry Controlled Other Party (ICOP), 152
Inflight Entertainment and Connectivity/Communication (IFEC), 180
In-flight entertainment (IFE) systems, 180
In-flight services, 93
Insurance costs, 35
Integrated vehicle health management (IVHM), 36, 102, 106, 107
Intellectual property (IP), 153, 158
Intergovernmental Panel on Climate Change, 134
International Aerospace Quality Group (IAQG), 152
International Air Transport Association (IATA), 1, 2, 85, 134, 136, 144
International Civil Aviation Organization (ICAO), 2, 17, 70, 102, 116, 128, 134, 138, 139, 143, 144
International Organization for Standardization (ISO), 152
Item Development Assurance Levels (IDALs), 125

Japan Airlines Boeing 777-200ER aircraft, 86
Joint Aviation Authorities (JAA), 70
Joint ventures, 87
Junkers Ju-52 "Iron Annie" monoplane, 45
Just in Time (JIT) approaches, 35

Kelly Act, 44

LaGuardia Airport, 91
Lancaster bomber, 13
Landing fees, 35
Latent failures, 117
LEAP model, 20
Leonardo da Vinci sketch, 10
Le Rhône 9C, 15
Lift-generating inclined plane, 10
Local air pollution, 30
Lockheed Electra, 45
Lockheed L-1011, 18
Lockheed Martin demonstrator project, 169
Lockheed P80 Shooting Star fighter aircraft, 16
Low cost carriers (LCCs), 26
Low Plasticity Burnishing (LPB®), 111
Low-pressure turbine (LPT), 167
Lufthansa Technik, 110, 111

MacDonnell Douglas DC-10, 18
Mach numbers, 65
Maintenance
  big barriers to entry, 107
  business models
    MROs, 107–109
    OEMs, 107–109
  costs, 35–36
  evolution, 106–107
  health monitoring, Big Data and analytics, 110
  ICAO "safety cycle," 102, 103
  programs and planning
    A-check, 104
    B-check, 104
    C-check, 104, 105
    D-check, 104, 105
    ETOPS, 106
    SB, 106
    service check, 104
  regulations, 103
  RFID, 110
  STC, 110–111
Maintenance planning document (MPD), 103, 104
Maintenance Program Industry Group (MPIG), 73
Maintenance, repair, and overhauls (MROs), 3–5, 95, 106, 107, 109, 181
  and airlines, 109
  OEMs, 108–109
Maintenance Review Board (MRB), 73, 74, 103
Maintenance Steering Group (MSG), 3, 18, 106
Maintenance Steering Group-3 (MSG-3), 103
Mammoth balloon, 10
Man-machine interface (MMI), 63
Manufacturing costs, 35
Martin aircraft, 13
Masks failures, 117
Meggitt Avionics, 10
Meteorologists, 92
Minimum Aviation System Performance Standards (MASPS), 150
Minimum Operational Performance Standards (MOPS), 150
Montgolfier, Jacques-Etienne, 10

National Aeronautics and Space Administration (NASA), 118, 162, 169, 171, 179
National Aerospace and Defense Contractors Accreditation (Nadcap) program, 151
Network planning, 83–85
NextGen, 138

Noise, 30, 142–144
Non-destructive testing, 59
North Atlantic Treaty Organization (NATO), 155
Northrop Corporation, 175

O'Hare Airport, 1
"On Aerial Navigation," 10
Oneworld, 36
Online Aerospace Supplier Information System (OASIS), 152
Open rotor engines, 34, 167
Open Skies model, 84
Operation, airline operators
- aircraft load agents, 90
- airport customer service, 89–90
- centers, 89
- corporate safety and security, 93
- dispatchers, 92
- diverted flight, 94
- flight operations, 93
- in-flight services, 93
- irregular operations, 94
- maintenance, 90
- meteorology, 92
- regulatory agencies, 88
- reservations/distribution channels, 89
- "reserve" crews, 94
- scheduling, 90–92
- security, 90
- servicing aircraft, 93
- supporting functions, 93–94
- weather, 94

Operators
- business strategy
  - carriers, 83
  - fleet strategy and products, 85–87
  - network planning, 83–85
  - perished inventory, 83
  - public service, 83
  - route networks, 83
- employees, 94–95
- finances
  - CASM, 87
  - "cost of capital," 87
  - costs, 88
  - fares and pricing, 88
  - long-term stability, 87
  - RASM, 87
  - U.S. carriers, 87
  - yield, 87
- high load factors, 82
- operation
  - aircraft load agents, 90
  - airport customer service, 89–90
  - centers, 89
  - corporate safety and security, 93
  - dispatchers, 92
  - diverted flight, 94
  - flight operations, 93
  - in-flight services, 93
  - irregular operations, 94
  - maintenance, 90
  - meteorology, 92
  - regulatory agencies, 88
  - reservations/distribution channels, 89
  - "reserve" crews, 94
  - scheduling, 90–92
  - security, 90
  - servicing aircraft, 93
  - supporting functions, 93–94
  - weather, 94
- passenger journey, 82
- passengers, 96–97

Original equipment manufacturers (OEMs), 3–5, 29, 30, 32, 35, 47, 54, 70, 71, 73, 75–77, 102, 103, 106–110, 152, 158
Ornithopters, 10
Overall pressure ratio (OPR), 169, 170

Paris Climate Agreement, 134
Particular risk analysis, 127
Parts Manufacture Authority (PMA), 108
Passengers, 96–97
Performance Review Institute (PRI), 151
Performance Review Institute Qualified Product List (PRI QPL), 152
Perry Memorandum, 119
Perry, William J., 155
Piccard, Bertrand, 140
Pipistrel Alpha Electro trainer, 172
Piston-engine airliners, 46
Planemakers
- digitally enabled manufacturing, 49–52
- expanding partnerships, 47–49
- jet age, 46–47
- lightweight metals, 42
- new industry
  - aerodynamic shapes, 45
  - Boeing Model 247, 45
  - civil aircraft, 45
  - commercial airplanes, 44
  - long-distance air flights, 44
  - mail-carrying services, 44
  - metal forming techniques, 45
  - monoplanes, 45
  - non-stop transatlantic crossing, 44
  - passenger flying boat service, 44
  - passenger service, 44
  - piston-engine civil transports, 46
  - powered airplane, 42
  - re-armament programs, 45
  - transcontinental air routes, 44
  - tri-motor airliner, 44
  - U.S. Air Mail service, 44
  - wire-braced biplane configuration, 44
- powerplant revolution, 53
- power-to-weight ratio, 41
- specialization grows, 42–43
- supply chain management, 53–55

Power-by-the-Hour™, 108
Powered flight, 58
Pratt & Whitney (P&W) Geared Turbofan™ (GTF™), 20, 165–167, 179
Preliminary aircraft safety assessment, 122, 123
Preliminary system safety assessment, 124–125
Pressurized civil airliners, 58
Preventative maintenance, 6
Production certificate (PC), 70, 71, 73
Prognostics and Risk Management, 110
Propeller-driven aircraft, 38
Propeller engines, 31
Propulsion
- AE, 171, 172
- Aerion AS2 aircraft, 170
- BLI/distortion-tolerant fan test, 171
- BPR, 166, 168–170
- CAFE, 171
- Concorde, 169, 170
- counter-rotating open-rotor test, 167
- distributed propulsion, 171
- electric propulsion, 171–173
- FPR, 170
- General Electric (GE) Aviation CFM56 engine core, 170
- global commercial airlines, 168
- HE, 171–173
- historical gas turbine efficiency *vs.* achievable capability, 166
- Jet A, 168
- LPT, 167

noise reduction, 172
OPR, 169, 170
Pipistrel Alpha Electro trainer, 172
Pratt & Whitney J58 variable-cycle engine, 170
P&W GTF, 166, 167
SAJFs, 168
superconductivity, 173
TSFC, 166
Public service, 83
P&W Wasp-powered Boeing 42, 13

Quality management systems, 152

Radio frequency identification (RFID), 110
Radio Technical Commission for Aeronautics (RTCA), 71, 118, 149–150
RAE thermal test rig, 62
Rapid And Cost-Effective Rotorcraft (RACER), 175
Reliability-centered maintenance (RCM), 18, 106
Research and development (R&D) investment, 44, 47
Reservations/distribution channels, 89
Reserve crews, 94
Revenue per available seat mile (RASM), 87
Reynolds numbers, 65
Risk and Revenue Sharing Partnerships, 4
Roaming agents, 90
Rolls-Royce, 16, 108, 109, 140, 179
Rolls-Royce Avon engines, 17
Roskill Commission, 143
Route networks, 83
Royal Aeronautical Society, 10
Royal Aircraft Establishment (RAE), 59
Ryanair, 87

Safety, 26, 33
and compliance
ICAO "safety cycle," 102, 103
maintenance evolution, 106–107
maintenance programs and planning, 103–106
regulations, 103
corporate, 93
and environmental requirements, 29–30
Safety assessments
ARP 4761, 119–120
design assurance, 118–119
DO-180, 118
origins of, 116–118
safety management and safety life cycle
continued airworthiness, 129
design and certification, 128, 129
operating environment, 128–130
production, 129
safety requirements development and validation, 120
aircraft functional hazard assessment, 121–123
preliminary aircraft safety assessment, 122, 123
preliminary system safety assessment, 124–125
system functional hazard assessment, 123–124
safety requirements verification, 120, 125–126
aircraft safety assessment, 127
common cause analyses, 126–127
system-level analyses, 126
system safety assessment, 127
software and hardware development processes, 125
Safety management system (SMS), 128–130
Scheduling, airline operators, 90–92
"Scope clause" provisions, 88
"Scope clauses," 86
Seats, 31
Security, 90
Service Bulletins (SB), 106
Servicing aircraft, 93
Sikorsky Aircraft Company, 14
Sikorsky X2, 174, 175
Single European Sky (SES), 138
Single European Sky ATM Research (SESAR), 138
SKYbrary, 73
Skyteam, 86
SkyWest Airlines, 86
Slots/takeoff and landing times, 84
Society of Automotive Engineers (SAE International), 71, 118
Solid-state electronic devices, 75
Sopwith Aviation Company, 13
Sopwith Camel biplane, 13
Southwest, 87
Southwest Airlines, 88
Spare aircraft, 91
Speed, 26
Standards
in civil aviation regulations, 155–157
and conformity, 151–152
dual-use standards and MilSpec reform, 154–155
early standards development organizations
ARINC, 149
ASD, 150
EUROCAE, 150
RTCA, 149–150
history and aerospace industry, 148–149
and innovation, 152–153
quality management systems, 152
SDOs, 153–154
use and content consumption, 157–158
value and drivers
aviation product, 151
global nature, 151
maintenance/replacement, 151
safety, 150–151
Standards and Recommended Practices, 116
Standards development organisations (SDOs), 153, 154, 157
Star Alliance, 86
Structural health monitoring (SHM), 107
Supersonic transport (SST), 18, 162
Supplemental structural inspections (SSIs), 106
Supplemental type certificate (STC), 70, 71, 110–111
Sustainable alternative jet fuels (SAJFs), 168
Sustainable Aviation Fuel Users Group (SAFUG), 141–142
System functional hazard assessment, 123–124
System-level analyses, 126
System safety assessment, 71, 127

Tampa Airboat Line, 44
Technical Standard Orders (TSO), 70–71, 156
Tentative Airworthiness Standards for Supersonic Transports, 117
"Terrorist-resistant" flight deck doors, 33
Tests
certification, 73–75
flight test, 64–66

new procedures, 62–64
performance monitoring, 65
static tests, 58–61
3D computer-controlled manufacturing tools, 58
Thermal shock tests, 61
Thomas Cook Airlines, 83
3D printing, 180
Thrust-specific fuel consumption (TSFC), 166
Ticket cost, 26
Top level requirements (TLRs), 30
Transatlantic flights, 84
Transportation Security Administration (TSA), 90
Transport Canada (TC), 103
"Tricycle" Landing gear arrangement, 32
TrueChoice™ Flight Hour, 108
TrueChoice™ Material, 108
TrueChoice™ Overhaul, 108
Truman, Harry, 166, 175
TSO authorization (TSOA), 71
Turbofan engine, 16, 19
Turboprop airliner, 46
Two-spool jet engine, 17
Type certificate (TC), 20, 33, 70
and PC, 71, 73
STC, 70, 71, 110–111

UAS Standardization Coordination, 154
UN Climate Conference in Copenhagen, 134
United Aircraft and Transportation Company (UA&TC), 13, 44
United Airlines, 14, 18, 44, 45, 83
Unmanned aircraft systems (UAS), 130
UPS, 85
U.S. carriers, 87
U.S. Department of Defense (DoD), 155
U.S. Federal Radio Commission, 149

Vertical takeoff and landing (VTOL), 162
Vickers Viscount, 46
Virgin Atlantic, 140
Visual flight rules, 138

Weather, 94
Whittle, Frank, 15, 16, 166
Whitworth, Joseph, 148
Widespread fatigue damage (WFD), 106
Wind tunnel testing, 33, 63, 64
World War I, 149
World War II, 175
Wright Flyer, 11, 12
Wright, Orville, 11, 12
Wright, Wilbur, 11, 12

Yield, 87

Zonal safety analysis, 126, 127
Zunum, 140

www.ingramcontent.com/pod-product-compliance
Lightning Source LLC
Chambersburg PA
CBHW041255150926
51501CB00011B/205
9780768093476